D0930606

HISTORY, PHILOSOPHY AND SOCIOLOGY OF SCIENCE

Classics, Staples and Precursors

HISTORY, PHILOSOPHY AND SOCIOLOGY OF SCIENCE

Classics, Staples and Precursors

Selected By

YEHUDA ELKANA
ROBERT K. MERTON
ARNOLD THACKRAY
HARRIET ZUCKERMAN

THE DISCOVERY
OF SPECIFIC
AND LATENT HEATS

BY

DOUGLAS McKIE

AND

NIELS H. DE V. HEATHCOTE

ARNO PRESS

A New York Times Company

New York – 1975

752767

Reprint Edition 1975 by Arno Press Inc.

Reprinted from a copy in
  The Princeton University Library

HISTORY, PHILOSOPHY AND SOCIOLOGY OF SCIENCE:
Classics, Staples and Precursors
ISBN for complete set: 0-405-06575-2
See last pages of this volume for titles.

Manufactured in the United States of America

————◆————

Library of Congress Cataloging in Publication Data

McKie, Douglas.
    The discovery of specific and latent heats.

    (History, philosophy, and sociology of science)
    Reprint of the 1935 ed. published by E. Arnold,
London.
    1. Specific heat--History. 2. Evaporation, Latent
heat of--History. 3. Black, Joseph, 1728-1799.
4. Wilcke, Johan Carl, 1732-1796. I. Heathcote, Niels
Hugh de Vaudrey, joint author. II. Title. III. Series.
QC295.M28   536'.63'09   74-26274
ISBN 0-405-06602-3

THE DISCOVERY
OF SPECIFIC
AND LATENT HEATS

BY

DOUGLAS McKIE, Ph.D., B.Sc.

LECTURER IN THE HISTORY AND METHODS OF SCIENCE,
UNIVERSITY COLLEGE, LONDON

AND

NIELS H. DE V. HEATHCOTE, B.Sc.

UNIVERSITY COLLEGE, LONDON

WITH A FOREWORD BY

E. N. DA C. ANDRADE, D.Sc., PH.D., F.R.S.

QUAIN PROFESSOR OF PHYSICS IN THE UNIVERSITY OF LONDON

EDWARD ARNOLD & CO.

LONDON

Printed in Great Britain by
Butler & Tanner Ltd., Frome and London

FOREWORD

At the present day it is fortunately not necessary to urge the interest and importance of a close study of the history of science. That this has been realized by our universities is sufficiently evidenced by the establishment, and subsequent success, of a school in the History, Methods and Principles of Science in the University of London, and by the recognition which the study and collection of the scientific instruments of bygone days has won in the University of Oxford. That the originating brains of science likewise acknowledge the lure of the historical aspect of their studies is well exemplified by the loving labour which Clerk Maxwell bestowed on editing the work of Henry Cavendish and, in more recent times, by the pious care devoted by Professor Dobell, the eminent protistologist, to his delightful volume on the work of Leeuwenhoek, not to mention such men as Mach and Duhem, who have become better known for their historical and philosophical studies than for their valuable researches in the sciences themselves. In my pleasant task of introducing the work of Dr. McKie and Mr. Heathcote I need, therefore, scarcely labour the general importance and significance of the work to which they contribute as historians of science.

Although this is the first independent book with which their names have been associated, both these writers are well known to those of us who are interested in the teaching of the history of science in the University of London. Dr. McKie is responsible for the instruction in the history of chemistry at University College, London, while both he and Mr. Heathcote have communicated careful studies on historical themes to *Science Progress*, a periodical which has always been prompt

3

to encourage an interest in the science of times past. They bring to their task mature judgment and sound knowledge of the general scientific background of the particular matter which is now their argument.

The subject with which the book deals, under its unpretentious title, is no less than the foundation of the modern science of heat, which may be said to have originated when a really clear distinction was made between heat and temperature. The importance of Black's work has always been recognized, but considerable obscurity has surrounded the services of his predecessors and contemporaries, notably Wilcke, in establishing the science. Robison, writing shortly after Black's death, did not scruple to suggest that the northern man of science had derived his work from a surreptitious knowledge of what Black had done. In this book the reader will find a very careful and objective discussion of the achievements of the Baltic workers, which appears to establish, among other things, that, while Wilcke's work was independent, there is no doubt that it was both later and less satisfactory than that of Black.

This study is distinguished by the clearness with which the somewhat confused and confusing views and notations of most of the early workers have been interpreted. Instead of dismissing as nonsense matter that at first reading appears as without significance, our authors have patiently applied their knowledge of contemporary modes of expression and thought until the writer's meaning, wrong though it may be, is revealed as self-consistent and comprehensible. Excellent and typical is what they have to say of Morin on page 64: "Perhaps we can appreciate the point of view of Morin's age if we consider *colour* instead of heat. Colour to us is a quality; we can imagine a scale in which blue, for example, is graduated in depth from very pale to very dark; by comparison with our scale we could define any given sample of blue as having so many 'degrees of blue'; we could say that 2 degrees of blue mixed with 4 degrees of blue would give a blue whose intensity was about 3; if we think of dyes, we could also argue that 2 measures of blue dye at 6 degrees would have twice the bluing effect of 1 measure —but it would surely never occur to us to try to estimate the *quantity of blueness* in a thing."

The book owes much of its excellence to the fact that throughout it is based on a study of the work of the original writers, in the original language. In the case of Wilcke, for instance, the Swedish text has been used. The authors have the expert knowledge and the patience needed for a thorough survey of the contemporary authorities. In these days of little thought and large books it is refreshing to find so much care and scholarship packed into so small a volume. It is with confidence that I recommend to the notice of those readers who like to think of science as a growing and organic thing this study, which I think may, without exaggeration, be called a model of its kind.

E. N. DA C. ANDRADE.

PREFACE

In presenting to the scientific reader this account of the discovery of specific and latent heats, we do not feel it necessary to give any of the customary excuses of authorship except to say that our researches constitute, so far as we are aware, the first detailed attempt at one of the most important chapters in the history of physics.

But it remains for us to express our thanks to those to whom we are indebted: to Professor E. N. da C. Andrade, D.Sc., F.R.S., Quain Professor of Physics in the University of London, for kindly contributing the Foreword; to the University of London for a grant in aid of publication; to Mr. W. A. Fleming, M.A., LL.B., J.P., Secretary to the University of Edinburgh, for providing us with information about William Cleghorn from the University records; to Professor T. S. Patterson, Ph.D., D.Sc., Gardiner Professor of Organic Chemistry in the University of Glasgow, and to the Librarians of the Universities of Edinburgh and Glasgow and of the City of Glasgow for assisting us in our regrettably unsuccessful search for the lost registers of the Literary and Philosophical Society of the University of Glasgow; to Miss Hilda Browning, Secretary of the Society for Cultural Relations with the U.S.S.R., and to MM. K. Sakharoff and M. Joukovsky of the Academy of Sciences, Leningrad, for their assistance in obtaining copies of the portraits of Krafft and Richmann; to the Director of the Synebrychoffska Gallery, Helsingfors, for kindly arranging to photograph the miniature of Gadolin, and to the Librarian of the Royal Swedish Academy of Sciences, Stockholm, for a similar kindness with regard to the portrait of Wilcke; to Messrs. Oliver and Boyd, Edinburgh, for permission to use the reproduction of Robison's portrait that recently appeared in the *University of Edinburgh Journal*; to the Secretary of the University of St. Andrews, the Registrar of the University of Glasgow, the Keeper of the Scottish National Portrait Gallery and the Director of the National Museum of Antiquities of Scotland for answering various inquiries; to the Royal Society of London for extending certain library facilities to us; and to the Library staffs of this College and of the British Museum for their able and willing assistance. Our acknowledgments for the permission of the owners to reproduce the various portraits are recorded on the reproductions.

Finally, we gladly acknowledge our great indebtedness to our former teacher, Mr. D. Orson Wood, M.Sc., F.Inst.P., Deputy Director of the Physics Laboratories of this College, for providing us with those initial references to the original literature that are invaluable to the researcher at the beginning of an investigation, for his enthusiastic and generous help throughout the course of our work, and for reading and criticizing our script before publication and for suggesting various corrections and amendments.

D. McK.

N. H. DE V. H.

UNIVERSITY COLLEGE, LONDON
April 1935

CONTENTS

LIST OF PLATES

THE DISCOVERY OF
SPECIFIC AND LATENT HEATS

CHAPTER I

THE WORK OF JOSEPH BLACK (1728–1799)

The discovery of specific and latent heats presents certain difficulties to the historian of science; and the assertion that Joseph
Black or, on the other hand, Johan Carl Wilcke was the founder
of the quantitative science of heat always involves the scrupulous
in some degree of doubt. These difficulties have arisen for
various reasons. The first authoritative account of Black's
researches was published posthumously forty years after they
were carried out, while Wilcke's work appeared in Swedish and
no complete or detailed account of it seems to have been generally available either to his contemporaries or hitherto, so far as
we are aware, to his successors.[1] The studies of both these
investigators will therefore be described below, and an attempt
will be made to clear away some of the historical obscurities that
have gathered around this subject.

Black's work was first published, four years after his death,
in his *Lectures on the Elements of Chemistry* (Edinburgh, 1803,

---

[1] German translations of the Transactions of the Swedish Academy
appeared in A. G. Kästner's *Abhandlungen* and *Neue Abhandlungen*, but
no use seems to have been made of them by Wilcke's contemporaries; a
French version of his paper on specific heat (1781) appeared in the *Observations sur la Physique* (1785, **26**, 256 and 381), and a summarized account
of the contents of this paper was included in Thomas Thomson's *System
of Chemistry* (Edin., 1802, I, pp. 309–14).

2 vols.),[1] edited by John Robison, who had been intimately associated with Black as pupil, colleague and friend from 1758 onwards and thus throughout the whole period covered by his researches on heat.   Robison was therefore admirably equipped for this task: he prepared his text from Black's lecture notes and from a manuscript of the lectures that had been copied by a student and subsequently corrected by Black when it came into his possession.

Black's Studies on " Capacity for Heat."   In the *Lectures* Black pointed out that our knowledge of the distribution of heat among a number of bodies in a system had been much advanced by the invention of the thermometer; that, when a hotter body was introduced into such a system, a thermometer applied to each of the bodies in turn after all action had ceased indicated the same temperature in each; and that the nature of this curious "equilibrium of heat" was not well understood until he devised a method for its investigation.   Boerhaave,[2] he said, had supposed that in this equilibrium there was finally an equal amount of heat in each equal volume of space irrespective of the nature of the body occupying that space, basing his conclusion on the fact that a thermometer registered the same temperature in each of the bodies composing the system.   This, argued Black, was confusing quantity of heat with intensity of heat, two things plainly different.   Moreover, if a cubic inch of iron and a cubic inch of wood, both at oven temperature, were held in the hand, the iron felt hotter than the wood and continued to feel hot longer than the wood did, while, if they were at snow temperature, the iron felt colder and continued to feel cold much longer; and therefore it could surely be supposed that in the first case the iron gave out more heat than the wood, and in the second that it absorbed more than the wood, and therefore that

---

[1] The subject of Heat is dealt with in Vol. I.   All references to that work will be indicated here by the page number only.

[2] *Elementa Chemiae*, Leyden, 1732, I, p. 188: English translation by Peter Shaw, *A New Method of Chemistry*, London, 1741, I, pp. 245–6. Hermann Boerhaave (1668–1738) was Professor of Chemistry, Botany and Medicine in the University of Leyden from 1709 onwards.

PLATE I.

Heath, sc. *Raeburn pnxt.*

JOSEPH BLACK (1728–99)

Entered University of Glasgow in 1746. Studied under Cullen. Graduated M.D. (Edin.) in 1754, his inaugural dissertation, *De Humore Acido a Cibis Orto, et Magnesia Alba*, subsequently expanded into *Experiments on Magnesia Alba, Quicklime, and some other Alcaline Substances* (Edin., 1756), which showed by rigorous quantitative methods that the mild and caustic alkalis owed their differences to the presence or absence of "fixed air" in their composition, becoming a classic of chemistry. Professor of Medicine and Lecturer on Chemistry in the University of Glasgow, 1756–66, succeeding Cullen, whom he again succeeded as Professor of Medicine and Chemistry in the University of Edinburgh, 1766–99. Friend and counsellor of James Watt.

"the iron has a greater capacity and attraction for heat, or the matter of heat, than the wood has" (p. 78).

It was [added Black] formerly a common supposition, that the quantities of heat required to increase the heat of different bodies by the same number of degrees, were directly in proportion to the quantity of matter in each; and therefore, when the bodies were of equal size, the quantities of heat were in proportion to their density. But very soon after I began to think on this subject, (anno 1760) I perceived that this opinion was a mistake, and that the quantities of heat which different kinds of matter must receive, to reduce them to an equilibrium with one another, or to raise their temperature by an equal number of degrees, are not in proportion to the quantity of matter in each, but in proportions widely different from this, and for which no general principle or reason can yet be assigned (p. 79).

By way of explaining the genesis of his views on the matter, he added that they arose from a consideration of certain passages in Boerhaave's *Elementa Chemiae* and Martine's experiments on the rates of heating and cooling of bodies.

Boerhaave described some experiments, possibly carried out by Fahrenheit,[1] in which equal quantities of hot and cold samples of the same liquid, water, vinegar, alcohol or oil, were mixed with the result that "they came to the same degree of heat, which was half the excess of the hotter quantity above the cooler." But when Fahrenheit, at Boerhaave's request, mixed equal bulks of water and mercury at different temperatures, he found that this simple law was no longer followed:

For if the water were hotter than the mercury, when equal bulks thereof were mixed, the degree of heat arising from the mixture, was always more than the half which was expected. On the contrary, if the quicksilver were hotter than the water, and equal bulks of each were intermixed, the temperature produced by them was always less hot than the half of the differences. And this diversity was always found, as if in the former case three parts of hot water had been

---

[1] D. G. Fahrenheit (1686–1736), who devised the well-known scale of temperature used in the thermometer now called after him. See also pp. 76-7 below for further comment on these experiments.

mixed with two parts of cold; or in the latter case, as if three parts of cold water had been mixed with two of hot: but when three equal bulks of mercury are taken, and two such bulks of water, it matters not whether you heat the mercury, or the water; since after mixtion the temperature will correspond to half the difference of the heat in each; as before in water, where equal portions were mixed (*op. cit.*, I, pp. 268–70: English trans., I, pp. 290–1).

Boerhaave concluded thence that heat was distributed in bodies "in proportion to their bulk, or extension, and not of their density" (*ibid*). Black found such a conclusion "surprising," because this very experiment contradicted it; and he argued that it showed that mercury "has less *capacity* for the matter of heat than water (if I may be allowed to use this expression) has; it requires a smaller quantity of it to raise its temperature by the same number of degrees" (p. 81).

Again, Martine,[1] in an essay composed in 1739 at St. Andrews, described the experiments that he had carried out to test the current opinion that the rates of heating and cooling of bodies varied inversely with their densities, an opinion that he suspected of being "a very plausible speculation." He found that "contrary to all our fine theory, Quicksilver, the most dense ordinary fluid in the world, excepting only melted Gold, is, however, the most ticklish next to Air; it heats and cools sooner than Water, Oil, or even rectified Spirit of Wine itself" (*Essays Medical and Philosophical*, London, 1740, p. 258).

Martine's principal experiment consisted in taking two thin-walled glass vessels of the same size and shape, one containing 15 ounces of mercury and the other an equal volume of water at the same temperature and each containing a sensitive thermometer. He then placed them close together "before a great fire at a considerable distance from it, but so that the heat should equally act upon them," and found that the temperature of the

---

[1] George Martine (1702–41), physician of St. Andrews. Educated St. Andrews, Edinburgh and Leyden. In a riot at St. Andrews in 1715, Martine rang the college bells to celebrate the proclamation of the Pretender. M.D. (Leyden), 1725. Physician to the army in America in 1740. Died 1741 in the expedition against Carthagena.

mercury rose much more rapidly than that of the water. In the converse experiment, the mercury cooled more rapidly than the water. Indeed, water, instead of gaining or losing heat 13 or 14 times faster than mercury, as current theory supposed, was about twice as slow as mercury in heating and cooling. Similar experiments with other liquids controverted the theory and led Martine to say that he knew "no stronger instance than this of the weakness, or, if I may venture to say so, of the presumptuousness of the human understanding, in pronouncing too hastily concerning the nature of things from some general preconceived theories" (*op. cit.*, p. 257).

Here again, according to Black, the observations were explicable in terms of "capacity for heat." "The quicksilver," he argued, "therefore, may be said to have less capacity for the matter of heat. And we are thus taught, that, in cases in which we may have occasion to investigate the capacity of different bodies for heat, we can learn it only by making experiments. Some have accordingly been made, both by myself and others" (p. 82). But he gave no details of his method of determining these "capacities for heat," and we have only Robison's account of it in the "Notes" appended to the *Lectures*.

Dr. Black [wrote Robison] estimated the capacities, by mixing the two bodies in equal masses, but of different temperatures; and then stated their capacities as inversely proportional to the changes of temperature of each by the mixture. Thus, a pound of gold, of the temperature 150°, being suddenly mixed with a pound of water, of the temperature 50°, raises it to 55° nearly: Therefore the capacity of gold is to that of an equal weight of water as 5 to 95, or as 1 to 19; for the gold loses 95°, and the water gains 5° (p. 506).

Subsequent improvements in calorimetric method will be described later (see pp. 122–30 below).

Black's Researches on Fluidity. Before Black's researches, it was considered that, when a solid had been heated up to its melting-point, it required only a small additional amount of heat to effect complete fusion, and conversely that, when a liquid had been cooled to its freezing-point, only a small diminution of its heat was necessary for complete solidification; and that in these changes there were no further additions or abstractions of heat than were recorded by a thermometer applied to the system.

This [said Black] was the universal opinion on this subject, so far as I know, when I began to read my lectures in the University of Glasgow, in the year 1757. But I soon found reason to object to it, as inconsistent with many remarkable facts, when attentively considered; and I endeavoured to show, that these facts are convincing proofs that fluidity is produced by heat in a very different manner (p. 116).

If the common opinion was correct, he argued, the large masses of snow and ice that accumulate in severe weather would be suddenly and completely converted into water as soon as the atmospheric temperature reached their melting-point, and the consequent torrents and inundations "would tear up and sweep away every thing, and that so suddenly, that mankind should have great difficulty to escape from their ravages" (p. 118). This does not happen: ice and snow melt only slowly, especially if they form large masses, and it takes many weeks of warm weather to dissolve them totally into water. Indeed, it is this same slowness in melting that allows of the preservation of ice during the summer in ice-houses and leads to snow remaining on many mountains through the whole summer.

These facts suggest that a great quantity of heat enters the melting ice during its fusion and that the long time required for this heat to be abstracted from the surroundings is the cause of the slowness of liquefaction. And other facts confirm it: ice and snow melt slowly in the air of a warm room; ice instantly seems to draw heat with great rapidity from the hand that holds it; a cold stream of air descends from a piece of ice suspended by a thread in the air of a warm room, its heat having been given to the ice and its density thereby decreased; and the only effect on the ice is to change it into water that is not in the least sensibly warmer than the ice was before, the heat that has entered being "absorbed and concealed within the water, so as not to be discoverable by the application of a thermometer" (p. 119).

Black therefore proposed to estimate the quantity of heat that entered the ice during fusion by determining the rates at which heat entered equal masses of ice and water when suspended in the air of a warm room.

In order [he says] to prepare for this experiment, I chose two thin globular glasses, four inches diameter, and very nearly of the same weight: I poured into one of them five ounces of pure water, and then set it in a mixture of snow and salt, that the water might be frozen into a small mass of ice. As soon as frozen, it was carried into a large empty hall, in which the air was not disturbed nor varied in its temperature during the progress of the experiment; and in this room the glass was supported, as it were, in mid air, by being set on a ring of strong wire, which had a tail issuing from the side of it five inches long, and the end of this tail was fixed in the most projecting part of a reading desk or pulpit: And in this situation the glass remained until the ice was completely melted.

When the ice was thus placed, I set up the other globular glass precisely in the same situation, and at the distance of 18 inches to one side, and into this I poured five ounces of water, previously cooled, as near to the coldness of melting ice as possible, viz. to 33 degrees, and suspended in it a very delicate thermometer, the bulb of which being in the centre of the water, and the tube being so placed, that without touching the thermometer, I could see the degree to which it pointed. I then began to observe the ascent of this thermometer, at proper intervals, in order to learn with what celerity the water received heat, stirring the water gently with the end of a feather about a minute before each observation. The heat of the air, examined at a little distance from the glasses was 47 degrees of Fahrenheit's scale.

The thermometer assumed the temperature of the water in less than half a minute, after which, the rise of it was observed every five or ten minutes, during half an hour. At the end of that time, the water was grown warmer than at first, by seven degrees; and the temperature of it had risen to the 40th degree of Fahrenheit's scale.

The glass with the ice was, when first taken out of the freezing mixture, four or five degrees colder than melting snow . . . but after some minutes, it had gained from the air those four or five degrees,

B

and was just beginning to melt, which point of time was then noted, and the glass left undisturbed ten hours and a half. At the end of this time, I found only a very small and spongy mass of the ice remaining unmelted, in the centre of the upper surface of the water, but this also was totally melted in a few minutes more; and, introducing the bulb of the thermometer into the water, near the sides of the glass, I found the water there was warmed to the 40th degree of Fahrenheit. . . .

It appears, therefore, from this experiment, that it was necessary that the glass with the ice receive heat from the air of the room during 21 half-hours, in order to melt the ice into water, and to heat that water to the 40th degree of Fahrenheit. During all this time, it was receiving the heat, or matter of heat, with the same celerity (very nearly) with which the water-glass received it, during the single half-hour in the first part of the experiment. . . . The whole quantity of heat, therefore, received by the ice-glass during the 21 half-hours, was 21 times the quantity received by the water-glass during the single half-hour. It was, therefore, a quantity of heat, which, had it been added to liquid water, would have made it warmer by 40 — 33 × 21, or 7 × 21, or 147 degrees. No part of this heat, however, appeared in the ice-water, except 8 degrees: the remaining 139 or 140 degrees had been absorbed by the melting ice, and were concealed in the water into which it was changed.

The communication of heat to the melting ice was easily perceived, during the whole of its exposition, by feeling the stream of cold air which descended from the glass (pp. 120–2).

Black regarded this method as an analysis of the atmospheric melting of ice and proceeded to a now more obvious method of quantitative study. He rapidly counterpoised a piece of dried ice laid on flannel and at melting temperature in the scale-pan of a balance by adding sand to the other scale-pan and at once plunged the ice into a nearly equal mass of warm water, the temperature of the water immediately before the addition of the ice being 190° F. The ice all melted in a few seconds and the temperature fell to 53° F. Neglecting quantities less than half a drachm and expressing all quantities in half-drachms, he found that the weight of the ice was 119, of the hot water 135, of the mixture 254, and of the glass of the containing vessel 16 half-drachms. By other experiments, he ascertained that the capacity for heat of 16 parts of hot glass was the same as that of 8 parts of equally hot water; and he therefore substituted in his calculation 8 half-drachms of warm water in place of 16 half-drachms of warm glass, which brought the effective quantity of warm water up to 143 half-drachms. The water was originally hotter than the ice by 158 degrees and, if the only effect that had been produced was that this heat had become equally distributed through the mixture, the temperature of the mixture should have been $(158 \times 143)/(143 + 119)$ or 86 degrees warmer than melting ice. But it was found to be only 21 degrees warmer; and therefore a quantity of heat had disappeared that would have made the mixture and the glass warmer by 65 degrees or would have increased the "heat" of a quantity of water equal in mass to the ice alone by $65 \times (143 + 119)/119$ or 143 degrees on Fahrenheit's scale (pp. 122-5).

The difference between this result and the previous one Black ascribed to experimental error; the inmost parts of the ice might not have reached melting temperature.[1]

A foot-note inserted here and of some interest to us reads: "These two experiments, and the reasoning which accompanies them, were read by me in the Philosophical Club, or Society of Professors and others in the University of Glasgow, in the year 1762, and have been described and explained in my lectures there, and in Edinburgh, every year since" (p. 125).

In some further experiments Black by adding a certain mass of ice caused the temperature of an equal mass of hot water to fall from 176° F. to the freezing-point,[2] and that of another similar amount, to which a little salt had been added, to fall from 166° or 170° F. to below the freezing-point, the fluid produced being colder than the ice was at the beginning, "a curious and puzzling thing to those unacquainted with the general fact" (p. 125).

---

[1] The mean of the two results obtained by Black is 141 degrees of Fahrenheit's scale, which is equivalent to 78 cals./gram, a value closely agreeing with that now accepted (79·74 cals./gram).

[2] Therefore, 176° − 32° = 144°, a result confirming the preceding ones.

"It is, therefore, proved," said Black, "that the phenomena which attend the melting of ice in different circumstances, are inconsistent with the common opinion which was established upon this subject, and that they support the one which I have proposed" (p. 125).

It appears that Black carried out quantitatively the converse of his first experiment, namely, the freezing of water exposed to cold air. "Many registers of it are to be found among his memorandums," wrote Robison (p. 127, footnote); but climatic conditions seldom allowed of the experiment being made, the cold not being constant enough for simple calculation. And it was in discussing this freezing of water that Black first introduced the term "latent heat" for the heat involved in liquefaction and solidification (p. 127). He also studied substances other than ice and applied his ideas to liquefaction generally, but he did not include details of these experiments in his lectures since his courses were directed to beginners. At his request, Irvine [1] carried out experiments which showed that the latent heat of fusion of spermaceti was 141 or 148 and of beeswax 175, these numbers indicating the number of degrees that a mass of each melted substance would be raised in temperature by the heat it absorbed in melting, while that of tin was 500, referred to the solid and not to the melted substance (p. 137).

Black's Researches on Vaporization. The ideas and methods that he had so successfully applied to the study of fluidity, Black shortly afterwards extended to that of vaporization, which he supposed likewise involved the slow entry of a considerable amount of heat not detectable by the thermometer. Here again, the commonly held opinion was that it was necessary to add only a small amount of heat to water at its boiling-point in order to convert it into steam; yet, if this were true, argued Black similarly as he had done in the case of liquefaction, water would boil and vaporize with an explosive violence equal to that of gunpowder. On the contrary, however, water did

---

[1] William Irvine (1743–87). Studied under Black at Glasgow, where he succeeded Robison as Lecturer on Chemistry in 1770.

not increase in temperature during boiling and much heat was required to convert it into vapour even when the boiling was violent; and a thermometer indicated that its vapour was no hotter than the boiling liquid. Black could

scarcely remember the time in which I had not some confused ideas of this disagreement of the fact with the common opinion. . . . But the importance of the surmise never struck me with due force, till after I had made my experiments on the melting of ice. The regular procedure in that case, and its similarity to what appears here, en-' couraged me to expect a similar regularity in the boiling of water, if my conjecture was well founded (p. 156).

He deferred making experiments for some time, because he thought it would be difficult, if not impossible, to obtain a steady source of heat. But, after a practical distiller had told him "that, when his furnace was in good order, he could tell to a pint, the quantity of liquor that he would get in an hour," Black found by some rough tests that small amounts of water evaporated in times closely proportional to their quantities even with a sensibly irregular fire; and, encouraged by this result, he proceeded to carry out quantitative experiments, described as follows:

Experiment 1st. I procured some cylindrical tin-plate vessels, about 4 or 5 inches diameter, and flat-bottomed. Putting a small quantity of water into them, of the temperature 50 degrees, I set them upon a red-hot kitchen table, that is, a cast-iron plate, having a furnace of burning fuel below it, taking care that the fire should be pretty regular. After four minutes, the water began sensibly to boil, and in twenty minutes more, it was all boiled off. This experiment was made 4th October 1762.

Experiment 2d. Two flat-bottomed vessels, like the former, were set on the iron plate, with eight ounces of water in each, of the temperature 50°. They both began to boil at the end of three and a half minutes, and in eighteen minutes more, all the water was boiled off.

Experiment 3d. The same vessels were again supplied with twelve ounces of water in each, also of the temperature 50°. Both began to boil at the end of six and a quarter minutes, and the water was all boiled off, from the one in twenty-eight minutes, and from the other in something more than twenty-nine.

I reasoned from these experiments in the following manner: The vessels in the first experiment received 162 degrees of heat in four minutes, or $40\frac{1}{2}$ degrees each minute. If we, therefore, suppose that the heat enters equally fast during the whole ebullition, we must suppose that 810 degrees of heat have been absorbed by the water, and are contained in its vapour. Since this vapour is no hotter than boiling water, the heat is contained in it in a latent state, if we consider it only as the cause of warmth. Its presence is sufficiently indicated, however, by the vaporous or expansive form which the water has now acquired.

In experiment second, the heat absorbed, and rendered latent, seems to be about 830.

In the third experiment, the heat absorbed seems to be somewhat less, viz. about 750. The time of rising to the boiling heat, in experiment third, has nearly the same proportion to that in experiment first, that the quantities of water have. The deficiency, therefore, in the heat absorbed, has been probably only apparent, and arising from irregularity in the fire. Upon the whole, the conformity of their results with my conjectures was sufficient to confirm me in my opinion of its justice. In the course of further experiments made both by myself and some friends, and in which the utmost care was taken to procure a perfect uniformity in the heat applied, the absorption was found extremely regular, and amounted at an average to about 810 degrees (pp. 157–8).[1]

In discussing this work and the analogous phenomena of liquefaction Black spoke as follows of the heat that he had thus detected: "In both cases, considered as the cause of warmth, we do not perceive its presence: it is concealed, or latent, and I gave it the name of LATENT HEAT" (p. 157).

Black went on to quote from his own observations and from those of others various qualitative confirmations of his views. He heated a corked phial half-filled with water to several degrees above boiling-point and then withdrew the cork. According to the received opinion, the whole of the water ought to have

[1] This value of 810 degrees corresponds to 450 cals./gram, the value now accepted being 539·1 cals./gram.

evaporated immediately; but, as he expected, only a portion of it was converted into vapour and, further, the temperature of the remainder fell to the ordinary boiling-point. Thus the excess of heat in the water converted only a part of it into vapour and the common opinion was wrong. He was anxious to obtain a quantitative result here to show the proportion between the heat that had disappeared and the quantity of vapour produced or water vaporized, but the vapour issued with such violence that it carried drops of water away with it. Watt,[1] he says, obtained a more satisfactory result by putting water to a depth of three inches in a small digester with the safety-valve open; and he found that, after the water boiled, one inch of water was vaporized in half an hour. Then he added water to make up for this loss and, repeating the experiment, closed the safety-valve as soon as boiling began and left the digester to heat for half an hour, the temperature rising many degrees above the boiling-point. Removing it from the fire, Watt opened the valve slightly and steam rushed out violently for two minutes; he then closed the valve and, after allowing the digester to cool, opened it and found that an inch of water had evaporated.

It is reasonable to conclude from this experiment [said Black] that nearly as much heat was expended during the blowing of the steam-pipe,[2] as had been formerly expended in boiling off the inch of water. For, before opening the valve, the temperature was many degrees above the boiling-point, and all this disappeared with the vapour. The same inference may be drawn from the time that the digester continued upon the fire with the valve shut, because we may conclude that the heat was entering nearly at the same rate during the whole time. It is plain, however, that the experiment is not of such a kind as to admit of nice calculation; but it is abundantly sufficient to show that a prodigious quantity of heat had escaped along with the particles of vapour produced from an inch of water. The water that remained could not be hotter than the boiling-point, nor could the vessel be hotter, otherwise it would have heated the water, and converted it into vapour. The heat, therefore, did not escape along

---

[1] James Watt (1736–1819), the great improver of the steam-engine.
[2] The pipe through which the steam issued from the digester.

with the vapour, but *in* it, probably united to every particle, as one of the ingredients of its vaporous constitution. And as ice, united with a certain quantity of heat, is water, so water, united with another quantity of heat, is steam or vapour (pp. 160–1).

To these Black added examples from the observations of Boyle and of Cullen. Boyle [1] found that hot water, well boiled to expel all air, began to boil on exhaustion in the receiver of his air-pump long before a good vacuum was produced, indeed after a few strokes of the pump, that the boiling continued violently as long as pumping continued and that the temperature of the water fell from hot to lukewarm (*New Experiments Physico-Mechanicall, etc.*, Oxford, 1660, p. 388). This, said Black, was an additional proof of the correctness of his theory, Boyle's observation having since been confirmed both by Robison and by Watt. Cullen [2] (*Essays and Observations, Physical and Literary, read before a Society in Edinburgh*, Edinburgh, 1756, II, p. 145) [3] had carried out experiments explicable in a similar way. He had been studying the heat and cold produced by mixing various bodies, liquids and solids. Dobson, who was making the experiments for him, observed that a thermometer dipped into spirit of wine at atmospheric temperature and then suspended near by in the air recorded a fall in temperature of two or three degrees. This reminded Cullen of some similar observations recorded by Mairan [4] (*Dissertation sur la Glace*, Paris, 1749, pp. 248 f.). Cullen then carried out further experiments with volatile liquids and found that a similar fall in temperature occurred, that the temperature continued to fall as long as the thermometer remained wetted with liquid and that the temperature could be made to sink further still by a fresh application of liquid. The rate of fall was accelerated by

[1] The Hon. Robert Boyle (1627–91).

[2] William Cullen (1710–90), Professor of Medicine at Glasgow and later at Edinburgh. Black studied under Cullen and subsequently succeeded him both at Glasgow and at Edinburgh.

[3] Cullen's paper was read on May 1st, 1755.

[4] J.-J. D. de Mairan (1678–1771). Author of works on the barometer and the aurora. Member of the Academy of Sciences, Paris, and Secretary after Fontenelle.

moving the thermometer to and fro in the air or blowing on its bulb with bellows. By such methods, he caused the temperature to fall from 44° F. to below the freezing-point in the case of spirits of wine and further still with other liquids. The phenomenon was best shown by means of an air thermometer. He concluded "that the cold produced is the effect of evaporation" (*loc. cit.*, p. 150). He repeated the experiments *in vacuo* to counter the notion that the evaporation of fluids depended on the solvent action of air and found that the same thing took place; and in one of these experiments made with nitrous ether, the atmospheric temperature being 53° F., he found that, on exhausting the receiver of the apparatus, the temperature had fallen so low that most of the water placed in a vessel surrounding that containing the nitrous ether was frozen and had formed a crust of ice on the vessel containing the ether. These observations also, said Black, were readily explicable by means of his theory.

Black then proceeded to apply his views to the process of distillation, remarking to his students: "All of you know well enough how the operation of common distillation is conducted" (p. 166); and in eighteenth-century Scotland, with its bitter resistance to the laws of excise, there is no doubt that they did know a great deal about it. He pointed out that the water in the refrigeratory of a still became so hot that it had to be replaced by a stream of cold water; and that fatal accidents had occurred when proper attention was not paid to this, owing to the inflammability of the uncondensed vapours. On October 9th, 1764, with the assistance of Irvine, he carried out at Glasgow a quantitative experiment described as follows:

Five measures (each containing 4 lb. 5 oz. and 6 dr. avoirdupois) of water, of the temperature 52°, were poured into a small still in the laboratory. The fire had been kindled about 40 minutes before, and was come to a clear and uniform state. The still was set into the furnace, and, in an hour and twenty minutes, the first drop came from the worm; and in three hours and forty-five minutes more, three measures of water were distilled, and the experiment ended. The refrigeratory contained 38 measures of water, of which the temperature, at the beginning of the experiment, was 52°. When one measure had come over, the water in the refrigeratory was at 76°. When two had come over, it was at 100°; and when three had come over, it was at 123°.

In this experiment, the heat, which emerged from three measures of water, had raised the temperature of the water in the refrigeratory from 52° to 123°, or 71°. Now 3 is to 38 as 71 to 899⅓, and the heat would have raised the three measures 899⅓ degrees in its temperature, if this had been possible without converting it into vapour. The heat of the vapour from which this emerged, was 212°, or 160° more than that of the water. Taking this from 899°, there remains 739°, the heat contained in the vapour in a latent state.

But this must be sensibly less than the truth. During the experiment, the vessels were very warm—the heat of the still as hot as boiling water, and the refrigeratory gradually rising from 52°, which was within a degree or two of the temperature of the air of the laboratory, to 123°. A very considerable portion of the latent heat of the steam must have been carried off by the air in contact with a considerable surface, some of which was exceedingly hot. A great deal must also have been carried off in the steam which arose very sensibly from the water in the refrigeratory, towards the end of the experiment. Mr. Irvin also observed, that, during the distillation, the temperature of the water which ran from the worm, was about 11° hotter than the water in the refrigeratory. The steam, therefore, at a medium, was not 160° hotter than the water, which ran from the worm, but 125°, its mean temperature being about 87°. This consideration alone will make the latent heat of the steam not less than 774 degrees, without any allowance for waste (pp. 171–3).

Further, calculating the value from the times occupied in reaching the boiling-point and the boiling off of the three measures, Black obtained a value of 750 (p. 173).

A few weeks later, says Black, Watt made some further experiments with a more suitable apparatus and obtained a value of 825. And Black adds that Watt subsequently informed him that, in his studies on the steam-engine, he carried out very careful experiments which showed that the heat that was rendered latent in the vapour and the heat that emerged from it later were exactly equal, and that "the heat obtainable from steam" at atmospheric pressure was not less than 900 and not more than 950 degrees of Fahrenheit's scale.[1] Black also proposed to obtain a further value by comparing the amounts of ice melted by boiling water and by steam, but Watt's experiments soon gave him results sufficiently accurate for his views. However, "the celebrated philosopher, Lavoisier,[2] took this other method of measurement, and invented a most ingenious apparatus for the purpose of measuring all productions of heat, by the quantity of ice melted by these operations of nature in which it appears" (p. 175). This does not read as if Black ever worked with an ice-calorimeter. Moreover, we find no evidence that he ever used the apparatus that appeared so regularly as "Black's ice-calorimeter" in text-books of thirty or so years ago and still occasionally appears nowadays in such works and elsewhere, e.g. in the article on Heat in the current edition of the *Encyclopædia Britannica*.

Later, Black described how Watt had found that the latent heat of steam increased when the water was evaporated *in vacuo*, a value of 1048 being obtained in a rough experiment; and how, in more careful experiments, Watt had found that the latent heat of steam from water boiling at 70° under diminished pressure approached 1300 and exceeded 1200 degrees, so that "there is no advantage to be expected, in the manufacture of ardent spirits, by distilling *in vacuo*" (p. 190).

Black's Attitude to Theory. Black was never a great lover of theory or speculation: indeed, he was given to suspect all theories and even spoke of them as "a mere waste of time and ingenuity" (p. vii). And he was so opposed to all hypotheses and systems of chemistry that he called his discourses, not "Lectures on Chemistry," but "Lectures on the Effects of Heat and Mixture" (p. lxiv). With regard to the opinion of Bacon [3]

---

[1] A very good result, approaching the modern value of 970 on this system of measurement.

[2] A. L. Lavoisier (1743–94), the famous chemist.

[3] Francis Bacon, Lord Verulam (1561–1626), in his *Novum Organum* (London, 1620), described a new method for increasing natural knowledge and, taking as an example the investigation of the nature of heat, induced from all the available facts and observations, classified according to certain

that heat was a mode of motion, Black pointed out that this had been adopted in two different ways; English philosophers considered that the motion inferred by Bacon was a motion of the small particles of bodies, whereas most French and German philosophers and Boerhaave thought that it was a motion, not of the particles of the body itself, but of a subtle, highly elastic penetrating fluid, pervading all bodies and contained in the pores between their particles.   Neither of these theories, said Black, fully explained all the known facts relating to heat (p. 33); and his own discoveries on "capacity" were totally inconsistent with the former theory, since, if that theory were correct, heat should be communicated according to the laws of motion, and the denser bodies would be the most powerful communicators of heat, whereas the reverse was frequently the case (p. 83).

In so far as Black favoured any theory of heat, he was inclined in his later years to prefer that put forward by William Cleghorn [1] in 1779, because it explained many of the most remarkable facts of heat.   A brief account of Cleghorn's views will therefore be of some interest here.

Cleghorn, in his inaugural dissertation for the degree of M.D. in the University of Edinburgh (*Disputatio Physica inauguralis Theoriam Ignis complectens, Edin.*, 1779), propounded a new theory of heat or fire, in which all thermal effects were explained by means of two principles.   Briefly, his views were as follows: The flow of heat or fire from hot to cooler bodies indicates that there is a repulsive force in the particles of fire, by which they mutually recede from one another —*jure concluditur, vim repulsivam particulis ignis inesse, qua mutuo inter se recedunt* (*op. cit.*, p. 16); the work of Black on "capacities for heat" had shown that there are scarcely two

rules, that heat was a mode of motion.   Cf. *Works*, ed. by J. Spedding, R. L. Ellis and D. D. Heath, London, 1858–9, I, pp. 236–62 and IV, pp. 127–50.

[1] Little is known of Cleghorn, except that he was the nephew of George Cleghorn (1716–89), the physician, that he was a student at Edinburgh from 1777 to 1779, and that he died a few years after graduating M.D. in 1779.   Apparently he studied under Cullen and Black; for both occupied chairs at Edinburgh during the years when Cleghorn was a student there.

bodies in which the quantity of fire is the same, and therefore there is in bodies a force attracting fire and differing in different bodies—*concludendum igitur est, corporibus inesse vim ignem attrahentem, et hanc esse in aliis aliam (ibid.,* p. 17); from these two observations, it follows that fire is distributed among bodies in direct proportion to their attraction for fire and in inverse proportion to the mutual repulsion between the particles of fire —*ex quibus principiis manifeste sequitur, ignem inter corpora distribui, ratione facultatis, qua ignem trahunt directe, et facultatis ignearum particularum repulsivae inverse (ibid.,* p. 17) [1]

Black went so far as to describe this theory as "the most probable of any that I know," but forthwith discounted his approval by adding that "it is, however, altogether a supposition" (p. 34).

In another place Black pointed out that the expansion of bodies by heat had led many to suppose that heat increased the weight of bodies, but that experiment showed the facts to be otherwise (p. 47); and added that, if heat depended on the addition of a subtle matter to bodies, then the matter "has not any perceptible degree of gravitation" (p. 48). Further, the observation that heat did not increase the weight of bodies was a powerful objection to Cleghorn's theory and, although it drove many philosophers back to Bacon's view, other studies tended to show "that heat is the effect of a peculiar substance" (p. 49); but his own discoveries made it difficult to avoid the conclusion that heat was a substance, although all speculations were hypothetical and to be avoided "as taking up time which may be better employed in learning more of the general laws of chemical operations" (pp. 192–3).

Black had repeatedly expressed the view that heat was the cause of fluidity and vaporization. Irvine put forward another view, namely, that it was not the combination of latent heat with solids and fluids that produced these changes, but that the real cause was that the capacity of a body for heat was greater when

---

[1] It is noteworthy that in this publication of 1779, the author praises Black's work and, referring to other workers in this field, says: *quibus omnibus facile praeripuit palmam Ill. Black, chemiae hic Professor* (p. 6).

the body was liquid than when it was solid; and that, for example, since ice had a less capacity for heat than water had, when ice was converted into water, a quantity of heat must enter the water without raising its temperature (see p. 133 below). To Black this was another way of saying the same thing; and the facts of supercooling[1] were against it, the evolution of heat when supercooled liquids solidified indicating that the retention of latent heat was the cause of their protracted fluidity (pp. 143 and 195).

[1] First observed by Fahrenheit (*Phil. Trans.*, 1724, **33**, 78).

CHAPTER II

DATE AND PRIORITY OF BLACK'S EXPERIMENTS

Robison states that it was between the years 1759 and 1763 that Black brought to maturity the speculations that had long occupied his thoughts on "the combination of heat or fire with the substance of tangible matter" (p. xxxiv); and adds that Black formulated during the summer of 1761 the design of comparing the times required to raise one pound of water one degree in temperature with the time required to melt one pound of ice, but there was at that time no ice-house in Glasgow and he had to wait until December before climatic conditions allowed him to carry out the experiment, which he made in a large hall adjoining his rooms in the University of Glasgow (p. xxxvi).

Black did not publish this work on the latent heat of fusion of ice, but he read an account of it on April 23rd, 1762, to a literary and philosophical society that met at the University, "as appears by the registers of the society" (p. xxxviii).[1]

Robison states too that Black had already made some notes on Fahrenheit's experiment [2] and drawn up plans for others before 1757, that some of these must have given rise to his theories of fluidity and vapour, and that before 1765 he had made numerous determinations "on the heats communicated to water by different solid bodies." These he discussed in his lectures at Glasgow and he was engaged on such experiments when called to Edinburgh in 1766. During this time, Irvine, subsequently his successor at Glasgow, assisted him in

---

[1] Efforts both by Professor T. S. Patterson and by us have unfortunately failed to trace these documents. See also p. 19 above.

[2] Cf. pp. 13–14 above.

this work and the registers of their experiments, found among Black's papers, most of them dated, and written partly by Black and partly by Irvine, were all previous to 1770 (p. 504). Clearly, Robison means "previous to the work of Wilcke," though he was in no necessity to concern himself about 1770, since Wilcke did not broach the subject of specific heat until his second paper which was published in 1781.

Watt, too, says Robison, assisted Black during this period and, during his first speculations on the steam engine, he was concerned with determining the capacities of various substances in order to make cylinders that required little heat to warm them and in which therefore little heat was wasted. But his invention of the separate condenser put an end to the need for these determinations.

It was also during this period of Black's tenure of the chair at Glasgow that Irvine made experiments on fluids that evolved heat on mixing, which led him to his method of determining the point of absolute privation of heat.[1] In an unreferenced quotation, which it is highly probable Robison took from the registers that he found among Black's papers, he states that Irvine's

fundamental proposition was that "the heat which appeared in mixing vitriolic acid and water is the difference between the sum of the absolute heats of the two ingredients, and the absolute heat of the mixture, while the heats, which each of them separately required for an equal variation of temperature, had the proportion of their respective absolute heats." Therefore, having discovered, by such experiments, the difference, and the ratio of the absolute heats of the ingredients, we can find their absolute heats, and the temperature at which those heats commence, or in which the ingredients contain no heat at all (p. 505).

After Black's departure to Edinburgh, Irvine continued this work on the absolute heats of bodies, and during this period a variety of terms was used both by Black and by Irvine in this field. Irvine's peculiar views led him to apply the term *affinity* and both he and Black used this. They also used

[1] See pp. 130–4 below.

PLATE II.

Raeburn pnxt.

JOHN ROBISON (1739–1805)

Graduated Glasgow, 1756. Served with the forces in Canada, accompanying Wolfe on his rounds to the posts on the river the night before the taking of the Heights of Abraham (1759). Succeeded Black as Lecturer on Chemistry in the University of Glasgow (1766). In Russia, 1770–3, as Professor of Mathematics to the St. Petersburg naval cadet corps: colonel and sometime inspector-general of the corps. Professor of Natural Philosophy in the University of Edinburgh (1773–1805) and first Secretary of the Royal Society of Edinburgh (1783–1805). "He was a man," wrote Watt, "of the clearest head and the most science of anybody I have ever known."

capacity, faculty for receiving heat, appetite for heat, and other terms; but at last they adopted *capacity* "as the best suited to all the occasional modifications of this property which appeared in the experiments" (p. 505).

Here Robison pointed out that this knowledge was familiar to students of chemistry at Glasgow and Edinburgh and that he had not the least doubt that certain Swedish students, several of whom were at Glasgow and Edinburgh between 1763 and 1770, carried this knowledge home with them; and these facts were important for settling the disputed claims of originality. Wilcke was clearly in his mind here, but, as appears below, his alarm was unnecessary and his suspicion ungrounded. He states, too, that Wilcke preferred the term *specific heat* and referred to the value obtained by comparison in bulk as *relative heat,* the latter being the product of *specific heat* and *specific gravity.* For example, the *relative heat* of gold is 1, the same as that of water; its *specific heat* is 0·0526 or $\frac{1}{19}$. Comparisons were usually made by bulk, but Wilcke, added Robison, thought that measures by weight were more proper "because the question seems to relate chiefly to the peculiar nature of the matter itself, independent of all consideration of space" (p. 507).[1]

According to Robison (p. xi), the experiment with ice and water immediately suggested to Black that similar thermal changes occurred in the conversion of water into steam. In fact, Black had suspected this from the outset on the grounds that liquids evaporated slowly; that steam was produced slowly, although the vessel containing the water was in contact with glowing fuel; that steam had a great scalding power; and that much heat was produced in the refrigeratory of a still. He had, indeed, been expressing this opinion in his lectures in 1761, before he had made a single experiment in this field. He spoke of the "vaporific combination of heat" (p. xli). From lecture notes taken in 1761, Robison was able to state that Black had already at that time "brought his thoughts on this subject to full maturity, and that nothing was wanting but a set of plain experiments, to ascertain the precise quantity of heat which was com-

[1] See pp. 106–7 below.

bined in steam, in a state not indicated by a thermometer, and therefore latent, in the same sense that the liquefying heat is latent in water" (p. xli).

But Black was more devoted to his university lectures and his patients than given to literary ambition; and moreover it was not until the summer of 1764 that he carried out any experiments that satisfied him as to the precise amount of heat latent in steam. Meanwhile, however, he had resolved in his mind the facts already mentioned and also the fact that the temperature of boiling water was constant; and he had verified all these facts by experiments and described them in his lectures

At this period Black was assisted by two of his most famous pupils, Watt and Irvine, who both carried out with their teacher extensive experiments in this field—"similar to those," wrote Robison, "which now fill many pages of the Memoires of the foreign Academies" (p. xliv). Robison could speak with authority here, since he was in possession of the register of experiments; and he urged Irvine's son to search through his father's papers with a view to publication (see below, p. 124).

In conjunction with his pupil Irvine, Black made his first experiments on "the latent heat of steam," finding that it was sufficient to raise 700–800 times as much water through one degree of temperature on the Fahrenheit scale, or, as Black put it, the steam contained 700 or 800 degrees of heat latent in it. In later experiments with Watt he found that the value was not less than 850 and sometimes greater, increasing as barometric pressure diminished (pp. xliv–xlv).

Black still left his work unpublished though it had enabled his friend Watt to carry out those remarkable improvements in the steam-engine that had already made him famous; he was more interested in conveying his ideas to his students. But Watt's engine became known throughout Europe; and it "attracted the attention of all those engaged in the great business of making money." "It was this," wrote Robison, "more than all the love of knowledge, so boldly claimed by the eighteenth century, that spread the knowledge of the doctrine of latent heat, and the name of Dr. Black" (p. xlvii); "Chemists, mechanicians, meteorologists, manufacturers," he added, "all took an interest in it; and publications and plagiarisms multiplied apace.   These were totally disregarded by the unambitious author, who, in the meanwhile, was happy in the success of his friend, and in the thoughts of having exerted his own talents so usefully for the public."   How far Robison's charge of plagiarism was justified will be seen below.   Black's friends repeatedly urged him to publish for the sake of establishing his claims of originality and priority, but his heavy duties, his ill-health and his fastidiousness in composition prevented him from going farther than forming a plan of the work.

It is clear from the above and from what has been said in the previous chapter that Black had formulated his doctrine of specific heat in 1760 and that of latent heat shortly after this; that he carried out the first determination of the latent heat of fusion of ice in December, 1761, and read an account of these experiments to the Literary and Philosophical Society of the University of Glasgow on April 23rd, 1762; and that he made the first determination of the latent heat of vaporization of water on October 4th, 1762, and the first determination of the latent heat of condensation of steam on October 9th, 1764, Irvine assisting him in the latter research.   And it will be shown below that he was the first investigator to formulate theoretical explanations of these phenomena and to test his theories by the necessary quantitative experiments.

Finally, there is one other important statement by Black with regard to the date of his studies on latent heat.   In a letter to Watt, dated 15th May, 1780, he wrote:

I began to give the doctrine of latent heat in my lectures at Glasgow, in the winter 1757–58, which I believe was the first winter of my lecturing there, or, if I did not give it that winter, I certainly gave it in 1758–59, and I have delivered it every year since that time in my winter lectures, which I continued to give at Glasgow until winter 1766–67, when I began to lecture in Edinburgh (J. P. Muirhead, *Correspondence of the late James Watt*, London, 1846, p. xxiii).

Black went on to refer to the fact that many distinguished foreigners had attended his early lectures and heard his explanations, added that about 1760–1 or soon after he read a paper on the subject to the "Philosophical or University Club" at Glasgow,[1] and concluded: "I could bring a multitude of other evidences to prove the early date of my doctrines on this subject."

The Problem of Black's Priority. In editing the *Lectures*, Robison in his annotations stressed the opinion that Black's priority in these discoveries was being filched from him by many later workers.

> These [he wrote, referring to Black's researches on the latent heat of fusion of ice] were most unquestionable inferences from observations the most familiar; and they now appear most obvious and easy: Yet, before Dr. Black, no person seems to have made them. Fahrenheit, Boerhaave, Mairan, de Luc, and all the inquisitive meteorologists of the two preceding centuries, though incessantly contemplating and employing the same facts in their disquisitions, never mention having had such a thought; nor is a trace of it to be seen in the laborious collections of that unwearied compiler, Professor Musschenbroeck.[2] It is the undivided property of my ingenious and acute preceptor (p. xxxvii).

Robison was correct in this; but he complained that Continental philosophers "seem agreed in giving the honours of the discovery of the unequal distribution of heat among bodies of different kinds to Professor Wilcke of Stockholm, who read a dissertation on it in the Royal Academy in 1771 (*Swed. Trans.*, vol. **35**, p. 93; *New Swed. Trans.*, 1782, Part Second)" (p. 504). Robison seems to have been unduly nervous in the matter and his statement about the continental philosophers appears to be exaggerated. His references were wrong, and he could not have been acquainted with Wilcke's memoirs. Wilcke's first paper on heat (1772, not 1771) dealt with the melting of snow, not with specific heat.

As for the continental philosophers, there is ample evidence

---

[1] On April 23rd, 1762, according to the Society's registers. See p. 31 above.

[2] P. van Musschenbroeck (1692–1761), Professor of Philosophy and Mathematics at Duisburg (1719), and later at Utrecht (1723). Professor of Philosophy at Leyden (1740). An extensive writer on physical science.

that at least some of them fully recognized Black's labours in this field. Landriani, for example, in a dissertation on latent heat, while noting that some authorities asserted that Wilcke had observed these phenomena prior to Black and had deduced similar consequences from them, ascribed the discovery of latent heat to Black (*Opuscoli Fisico-Chimici*, Milan, 1781, p. 100.[1] Pictet (*Essais de Physique*, Geneva, 1790, I, p. 23: English trans. by W. B., *Essay on Fire*, London, 1791, pp. 34–5) [2] also ascribed the discovery of latent heat to Black; and earlier in this work, in discussing specific heat, he stated that he found the first traces of the notion in de Luc's *Recherches sur les Modifications de l'Atmosphère* (Geneva, 1772, 2 vols.),[3] "and it had been already developed in the lectures given at that period by Dr. Black, of Edinburgh" (*op. cit.*, pp. 16–17: English trans., pp. 24–5).

But Séguin [4] in an extensive critical review of the contemporary knowledge of heat (*Ann. de Chim.*, 1789, 3, 148–242) [5] gave Crawford, Lavoisier and Laplace credit for the advances in specific heat without any mention of Black, to whom, however, he ascribed the discovery of latent heats of fusion and vaporization, stating incorrectly that the discovery was made in 1756. He added that on the Reaumur scale Black's value for

---

[1] A translation of the dissertation on latent heat appeared in the *Observations sur la Physique* (1785, 26, 88 and 197). Landriani's experiments provided qualitative confirmation of Black's theories from a study of the solidification of sulphur and certain alloys.

[2] M. A. Pictet (1752–1825). Succeeded de Saussure in chair of philosophy at Geneva in 1786. Only Vol. I of the *Essais* was published.

[3] " Je ne sais pas par exemple si nous nous faisons une juste idée de ce que c'est que l'*égalité* ou la *différence* de *chaleur* dans des corps de différente nature, dès que nous voulons pénétrer au delà des apparences, ou des indications du Thermomètre. Il est très peu probable que des corps différens, que nous disons *également chauds*, parce qu'ils tiennent le Thermomètre au même *degré*, contiennent la même quantité de *feu*, sous le même volume, ou même dans ses masses égales " (*Recherches, etc.*, II, p. 351).

[4] Armand Séguin (1765–1835) was from 1800 onwards one of the editors of the *Annales de Chimie*.

[5] The second part of this long memoir, a part that has, however, no interest for us here, appeared in *Ann. de Chim.*, 1790, 5, 191–271.

the latent heat of fusion of ice was 62·22°, Lavoisier and Laplace's 60° and Wilcke's 57·73° (footnote, p. 171), but merely quoted Wilcke's result without giving details of his method; and he further stated that Wilcke had used the method of the ice-calorimeter [1] before Lavoisier and Laplace (p. 235), although the latter were unaware of Wilcke's work when they published their memoir and had in fact constructed an apparatus surmounting the difficulties that had baffled Wilcke.

There is, however, another side to this question. In the absence of any publication by Black and through the dependence of continental philosophers for accounts of his researches on works emanating from English sources, any misunderstanding that may have arisen can be fairly blamed on Crawford,[2] whose *Experiments and Observations on Animal Heat*, etc. (London, 1779: 2nd edn., London, 1788), contained many misleading statements. This work, highly personal in tone, gave detailed accounts of the author's own researches, and the unwary reader, remote from the scene of events, might well imagine on perusing Crawford's pages that he was being introduced, if not to one of the founders of the quantitative science of heat, at least to one of its extensive improvers. It certainly misled Séguin, whom we

[1] The material theory of heat is generally referred to nowadays as the caloric theory, although this latter term was not introduced until late in its history. The term *caloric* was coined in 1787 to replace *matter of heat* by the French chemists, Guyton de Morveau, Lavoisier, Berthollet and de Fourcroy, in their revision of chemical nomenclature and it first appeared in their work, *Méthode de Nomenclature Chimique* (Paris, 1787, pp. 30–1; English trans. by J. St. John, *Method of Chymical Nomenclature*, London, 1788, pp. 22–3). The term *calorimeter* dates from 1789 and was coined by Lavoisier (*Traité Elémentaire de Chimie*, Paris, 1789, II, pp. 389–90; English trans. by R. Kerr, *Elements of Chemistry*, Edin., 1790, p. 345). The earliest references to *caloric* and *calorimeter* given in the *Oxford English Dictionary* are respectively 1791 (from Erasmus Darwin, *Botanic Garden*) and 1794 (from G. Adams, *Natural and Experimental Philosophy*), but the translations named above precede these and indicate the origin of both terms.

[2] Adair Crawford (1748–95), physician at St. Thomas's Hospital, London, and later Professor of Chemistry at the Royal Military Academy, Woolwich. Visited Scotland in 1776 and began his experiments on heat at Glasgow in 1777.

have just mentioned, and Magellan, who will be dealt with presently. Séguin, in fact, gave Crawford great place in these matters; and he admired this author to such a degree that some sections of the review to which we have referred are (unacknowledged) word-for-word translations of paragraphs from his book.

Crawford's manner of giving Black credit for his work on specific heat was peculiar. He indicated in a footnote (1779 edn., p. 12: 1788 edn., p. 88) that the facts had been taught publicly by Black and Irvine in the Universities of Edinburgh and Glasgow "for several years."[1] In the first edition this footnote expressed the hope that "those learned and ingenious philosophers will soon favour the world with their respective discoveries in this branch of science"; in the second Crawford appeared to have abandoned his hope, since he omitted this part of the note. Moreover, in both editions Crawford stated in footnotes that he left to Black and Irvine "the full and complete illustration" of the doctrine of specific heat (1779 edn., p. 15: 1788 edn., p. 92). The first edition gave no account of latent heat, except a very indefinite sort of recognition that Black had observed in a great variety of phenomena that cold was produced by the evaporation of water and heat by the condensation of vapour (p. 47) and no measurements were mentioned, but an acknowledgment was made of the helpfulness of Irvine's explanation of "Dr. Black's celebrated discovery of latent heat" (footnote, p. 17); the second edition did not include this, but explained that

this important fact was observed by Mr. de Luc, in the years 1755 and 1756. . . . It appears to have been discovered nearly at the same time by Dr. Black of Edinburgh, who had no opportunity of becoming acquainted with what Mr. Luc had done, and who taught it publicly in his Chemical Lectures, as early as the year 1757 or 1758. It is proper to add, that Dr. Black was the first who compleatly established this discovery by a series of ingenious and decisive experiments (pp. 71-2).

---

[1] Crawford states that he attended Irvine's lectures at Glasgow (1779 edn., p. 17, footnote). This probably occurred during his visit in 1776 (see footnote 2 on p. 38 above).

This statement was immediately followed by an account of Black's experiments on the melting of ice that appears to be wholly imaginary (pp. 72-4). A footnote (p. 72) stated that "a very accurate account of the Doctrine of Latent Heat has been also published by Mr. Wilkie (*sic*) in the Swedish Transactions. See Acta Suec. 1772, p. 97." Crawford could not have read this,[1] which, as shown below, was far from "a very accurate account of the Doctrine of Latent Heat." Crawford asserted too (p. 77) that Black was first led to the discovery of the latent heat of vaporization by the unexpected result of an experiment with Papin's digester. This statement was absurd; as we have seen above (pp. 23-4), Watt carried out this experiment for Black as a qualitative test of the latter's theory, and the result was as predicted. Crawford was a muddled and careless writer on matters of history; and he was inconsistent, since we find him towards the end of the book (1788 edn., p. 483) crediting Black with the discovery of both specific and latent heats. There can be, however, little doubt that his book brought him great reputation in France, where he was regarded as an authority in these matters.[2]

Indeed, Crawford's reputation in this field can be gauged by the recognition given him by Magellan [3] in his *Essai sur la Nouvelle Théorie du Feu Élémentaire, et de la Chaleur des Corps* (London, 1780). This author discussed the contemporary

[1] This is confirmed by a footnote on p. 431, which states that "Wilkie" in his "Essay on Latent Heat" (presumably this means the memoir of 1772, although it was not an essay on latent heat) had observed that the absorption and extrication of heat in the melting and freezing of bodies tended to retard the progress of these processes, and that Black taught the same doctrine. We know that Black held this view; it was one of the fundamentals of his theory and one of the steps by which he was led to his great discoveries. But Wilcke, as appears below, had no such clear ideas of the matter, and Crawford's statement is wrong.

[2] Du Carla in certain meteorological studies referred to Crawford as having proved that there was more latent heat in water than in ice and in steam than in water (*Observations sur la Physique*, 1783, **22**, 433).

[3] J. H. de Magellan (1723-90), lineal descendant of the famous Portuguese navigator. Abandoned monastic for scientific life. Came to England in 1764 and settled in London about 1778.

advances in the knowledge of heat and bestowed high praise on Crawford, to whose publication (*Animal Heat*, 1779) he ascribed the rise of a new branch of physics—"C'est à la publication de l'excellent Ouvrage du Dr. Adair Crawford . . . qu'on doit la naissance de cette branche de Physique" (*Essai*, p. 165). And he referred to the "découverte heureuse du Dr. Black, Professeur de Chimie à Edimbourg (ou plutôt de Mr. Wilcke, Professeur de Physique à Stockholm)" as "le germe de la théorie lumineuse, que le Dr. Crawford présenta au Public sur ce sujet" (*ibid.*). The discovery Magellan had in mind was, of course, that of latent heat; and he ascribed it entirely to Wilcke on the grounds that Wilcke had published his researches,[1] whereas Black had not—"C'est à ceux, qui publient leurs propres découvertes, & même celles des autres, que le Public en est redevable" (*ibid.*, p. 166). These were not the only mistakes that Magellan made here; he told his readers that Crawford's book had been so well received that it had been sold out in a few months and that a second edition was in rapid preparation, whereas in fact there were so many errors in the first edition that the second involved extensive corrections and consequently did not appear until 1788. Moreover, Magellan himself stated that several correspondents had complained to him that they found difficulty in grasping the principles set forth, admittedly very cursorily, in Crawford's book, and hence the appearance of the *Essai*—" C'est, peut-être, parcequ'il n'a pas mis ses idées dans un plus grand détail; ni les a-t-il assez accommodées à la portée de tout le monde." But they were not Crawford's principles and ideas: they were Black's; and Crawford himself, as we have seen, avoided the full illustration of

---

[1] Although the only reference he gave (and this undetailed) was to the merest mention of Wilcke's work in Bergman's *Opuscula Physica et Chemica* (Stockholm, 1779, I, p. 235). Cf. the English version (*Physical and Chemical Essays*, London, 1784, I, pp. 286-7), where Wilcke's name is omitted, although reference is made to the absorption of 72 "degrees" of heat by snow in melting, the figure 72 appearing here presumably on account of what is stated in the *Preface* (p. vii), namely, "because the degrees upon the Swedish thermometer are not reducible to integral numbers upon that of Fahrenheit."

them for obvious reasons. Besides, these ideas had already been set forth in detail and "assez accommodées à la portée de tout le monde" by Black in his chemical lectures at Glasgow and Edinburgh for nearly twenty years.

Magellan wrote the *Essai*, he said, to inform his friends about the new doctrines of heat. He did not arrive in England until 1764 and, although he was unacquainted with the previous history of this work, he appears to have assumed an unwarranted authority in these matters. His attitude to Watt makes this evident; for, to tell the great improver of the steam-engine that the public was indebted to those who *published* their discoveries must have seemed a quaint doctrine to that patient thinker, as he recollected the connection between Black's discoveries and his own thoughts on that famous Sunday afternoon stroll on Glasgow Green in the spring of 1765, when he hit upon the idea of the separate condenser as an improvement for Newcomen's engine.[1]

It is in this *Essai* that, so far as we are aware, the term *chaleur spécifique* or specific heat was first used, but not in its modern sense, since it was here taken to be the total heat in unit mass of a body in a certain specified state. Magellan noted that, according to Black's pupils, the difference between the *chaleur spécifique* of a body in the solid state and that of the same body in the fluid state was called *chaleur latente*; and he objected to the use of such a term as savouring of the occult and resembling too much the language of the Peripatetics (p. 173). However, he quoted Black's values of 147 and 900 degrees of Fahrenheit's scale as the respective latent heats of fusion of ice and vaporization of water; as for the former, he noted that Wilcke had found a value of 130 degrees, and as for the latter, he wrote:

On assure qu'il y a eu des expériences faites en Ecosse, qui determinent ce fait intéressant: & qu'on les a annoncées en quelques *Cours de Chimie*. Mais ceux qui en parlent, ne sont point d'accord dans leurs raports. Il y en a un, qui, plus zélé que son maître pour l'avancement des connoissances humaines, publia un *Essai* sur ce sujet,

---

[1] Cf. Hart, *Trans. Glasgow Archæol. Soc.*, 1859, **1**, 1.

où il declare, que la *vapeur* de l'eau rarement est plus chaude que *l'eau bouillante*; quoiqu'il y ait 790° de *chaleur latente*. Voyez *An Inquiry into the Effects of Heat*, London, 1770, in 8vo, *page* 48 & 49. Un autre plus moderne, le Dr. Leslie,[1] assure, d'après les calculs de deux Professeurs célébres d'Ecosse, que la *chaleur latente*, ou, selon mes idées, la *chaleur spécifique* de la *vapeur de l'eau* va jusqu'à 800° (pp. 173–4).

But he did not know whether there were any direct and decisive experimental proofs of this statement; if he had properly read the anonymous *Enquiry* of 1770, to which he referred, he would have found such proofs.[2]

The method described by Magellan for the determination of specific heats was identical with that already given by Crawford (see pp. 126–7 below) with the additional correction for the thermal capacity of the thermometer that was suggested by Richmann's investigations (see pp. 68–9 below). Magellan further included a table of specific heats as determined by Kirwan[3] and communicated to him (*Essai*, p. 177). This was the first published table of specific heats.[4] Kirwan's method

---

[1] P. D. Leslie, *A Philosophical Inquiry into the Cause of Animal Heat*, London, 1778, pp. 319–20. The "deux Professeurs célébres d'Ecosse" were Black and Irvine; and Leslie gave Black credit for discovering latent heats both of fusion and of vaporization, quoting Black's values for these as 147 degrees (pp. 313–14) and 800 degrees of Fahrenheit's scale respectively.

[2] On pp. 48 and 49, i.e. the pages to which he had just referred. This publication is discussed below (pp. 50–2).

[3] Richard Kirwan (1733–1812), chemist and geologist.

[4] In this table of specific heats, that of water was taken as = 1. The value for ice was 0·9, for mercury 0·033, for iron 0·125, for tin 0·068, for lead 0·050, for dephlogisticated air (oxygen) 87·000, for atmospheric air 18·670 and for fixed air (carbon dioxide) 0·270. Altogether this table gives 42 specific heats. But they are not all due to Kirwan; for we find by calculation that twelve of them, including the three for gases, have been derived from data obtained by Crawford and published in his *Animal Heat* (1779). Crawford was the first to attempt to determine the specific heats of gases. His method consisted of heating the gases to known temperatures in bladders and then immersing the bladders in water and noting the rise of temperature: essentially it was the method of mixture. But it was subject to such relatively large corrections and the increments of temperature in the water were so small that, even when Crawford later on

as given here was much the same except that the correction for the loss of heat in cooling was made by applying either Newton's Law of Cooling or Richmann's formula [1] (*ibid.*, p. 179).

Before Magellan went to press with his *Essai*, he sent proofs to Watt and received replies from the latter relating to Black's work and to other matters, such as Watt's own work and also Irvine's. These Magellan referred to in "Additions et Corrections" and in a "Post-Scriptum" (p. 192). Watt had written to point out that Black had discovered latent heat before 1758 or possibly 1757 and that he (Black) was going to publish his work "this summer," but Magellan declined to withdraw his commendation of Wilcke or to recognize Black's work, adding that "le Public ne peut pas manquer d'attendre avec empressement, grands efforts de génie & connoissances très importantes, dans cette publication du Dr. Black; puisqu'il a eu pas moins de 22 à 23 ans, pour l'enricher, & pour la perfectioner" (p. 192).

It is little wonder that Robison (p. 524) took exception to Magellan's "very authoritative manner."

Reverting to Robison's criticisms, we cannot avoid the conclusion that there must have been many who regarded Lavoisier as the discoverer of latent heat and this appears, as Robison suggested (p. 523), to be almost certainly due to Lavoisier's "neglect" to mention Black in the memoir that he published jointly with Laplace, describing the well-known ice-calorimeter and announcing the results of the researches in which it

(*Animal Heat*, 1788 edn.) used thin brass cylinders in place of bladders and oil instead of water, the results obtained were still far from satisfactory. The specific heats of inflammable air (hydrogen), dephlogisticated air, atmospheric air, fixed air and phlogisticated air (nitrogen) given in the 1788 edition (p. 489) are 21·4, 4·749, 1·79, 1·0454 and 0·7936 respectively.

[1] In Richmann's formula, if a is the difference of temperature between the cooling body and the air, and b the decrease of temperature in time interval t, the difference of temperature between the cooling body and the air after n intervals of time t is $\dfrac{(a-b)n}{a^{n-1}}$: and the amount of cooling in the nth time-interval is $b\dfrac{(a-b)^{n-1}}{a^{n-1}}$ (*Nov. Comment. Acad. Sci. Imp. Petrop.*, 1747–8, **1**, 193).

was employed (*Mém. Acad. R. Sci.*, 1780, p. 355).[1] It is impossible to conclude from this memoir that Lavoisier was unacquainted with the fact that Black had carried out pioneer work in this field, since it refers to the first edition of Crawford's *Animal Heat* (1779); and mention of Kirwan's figure for the specific heat of ice suggests that Lavoisier had also read Magellan's *Essai*.[2] The memoir was included in the *Mémoires* of the Academy for 1780, published in 1784.[3] Meanwhile, according to a footnote (p. 373) the authors had for the first time become acquainted through the Stockholm transactions for 1781 with Wilcke's work on specific heat, a recognition of another worker that makes their ignoring of Black appear even more deliberate.

There is however much more convincing evidence that Lavoisier was familiar with Black's work before his own researches of 1780. It appears that Desmarets had described to the Academy the experiments carried out by Black on the passage of water into ice and of ice into water, and that Lavoisier reported to the Academy his own observation of a fact of the same nature, namely, that the temperature of a mixture of ice

---

[1] It is unnecessary to give details of this work here, as it is well known and is nowadays readily available in the reprint published in 1920 by Messrs. Gauthier-Villars & Cie, Paris, in their series *Les Maîtres de la Pensée Scientifique*.

[2] The *Essai* with certain minor alterations was reprinted in the *Observations sur la Physique* (1781, **17**, 375 and 411), i.e. well before the publication of Lavoisier and Laplace's memoir. Lavoisier was a contributor to this journal.

[3] The reason for the inclusion of this paper, dated June 18, 1783, in the *Mémoires* for 1780 (published 1784) is not apparent. Lavoisier's laboratory note-books for the period 1780–1 contain entries relating to only three experiments on heat, namely, the latent heat of vaporization of ether and the latent heats of fusion of wax and tallow: and they show that it was not until the winter of 1782–3 that he collaborated with Laplace in studying the phenomena of heat by means of the ice-calorimeter. The work was continued in the winter of 1783–4. These facts have already been quoted from Lavoisier's note-books by Berthelot (*La Révolution Chimique— Lavoisier*, Paris, 1890, pp. 279–80, 283–6 and 294–5), and there does not seem to be any reason why a date earlier than the winter of 1782–3 should continue to be ascribed to this work.

and water did not rise until the last fragment of ice was melted (*Œuvres de Lavoisier*, Paris, 1864–93, V, p. 241). The full account of this incident is to be found in the early numbers of the journal edited by the Abbé Rozier. Volume I of this periodical appeared in 1773 as the *Observations sur la Physique*, but the various monthly numbers that had previously appeared in 1772 and part of 1771 had become scarce and were reprinted in two volumes in 1777 as the *Introduction aux Observations sur la Physique*.[1] The observation by Lavoisier to which we have just referred appears in the *Introduction* (2, 510) under Lavoisier's name, the communication of Desmarets being mentioned; and an anonymous account [2] of Black's researches appears earlier in the same volume (2, 428), but no reference is made to Desmarets.

This paper on Black's work was published in the number for September, 1772. "L'école d'Edimbourg," states the anonymous contributor, "fera époque dans l'Histoire de la Physique, pour s'être principalement occupée du feu & de l'air fixe. Le Docteur Black est celui qui a le plus multiplié les expériences sur ces deux points importants." The contents of the paper are limited to brief descriptions of six experiments with summary explanations. The first experiment showed that the temperature of a heated mixture of ice and water remained constant at F.P., as long as any of the ice was unmelted, but that it rose as soon as fusion was complete. It was concluded thence that ice continued to receive heat in melting and that this heat was rendered insensible in the absorption, since the thermometer registered a constant temperature. The second experiment was the converse of this, dealing with the freezing of water; and the explanation ran in accordance with that of the first experiment. The third experiment showed that, when an aqueous solution of salt was cooled, its temperature fell below the F.P. of water to the F.P. proper to such a solution and remained con-

---

[1] The reprinted *Introduction* only has been accessible to us.

[2] Described as having been sent from Edinburgh by one of Black's pupils (French, *disciples*). Possibly this is the substance of Desmaret's communication to the Academy.

stant at that point during the process of freezing. In the fourth experiment it was shown that supercooled water remained fluid below 32° F. and that its temperature, when solidification took place, rose instantly to 32° F., whence it was concluded that what retarded freezing also retarded the dissipation of heat and therefore that loss of heat was necessary for freezing. The fifth experiment demonstrated that, while the temperature of a mixture of equal masses of water at different temperatures was the mean of these temperatures, the temperature of a mixture of equal masses of snow and water was below the mean of the original temperatures; and therefore it again appeared that heat was absorbed in the process of melting.

From these five experiments it was concluded that all bodies in passing from the solid to the liquid state absorbed a certain amount of heat and that their fluidity was attributable to this absorption. Conversely, liquids on solidification gave out the *chaleur cachée* that was the cause of their fluidity. Black, it was added, had also shown that water on conversion into steam absorbed "une quantité prodigieuse de chaleur"; and, while steam was no hotter than boiling water, the steam from a still gave more heat to the water in the refrigeratory than did the same amount of boiling water.

The sixth experiment described the method of determining the heat absorbed in the melting of snow. It read:

Mettez dans un lieu chaud, une quantité d'eau nouvellement dégelée, & une quantité de neige prête à se dégeler, l'eau acquerra sensiblement tant de degrés de chaleur par heure, ou par minute, tandis que la neige ne paroîtra en recevoir aucun jusqu' à ce qu'elle soit fondu. Cette expérience est infaillible; & en fixant la quantité de chaleur reçue par l'eau dans un tems donné, on peut calculer la quantité absorbée pour produire la fluidité.

This ended the account. No values were quoted, but this sixth experiment gave the necessary details of Black's first method. Whether or no this is a summary of the communication made by Desmarets to the Academy does not concern us here, but it is established beyond all doubt that in 1772 Lavoisier was already familiar with Black's theory and method. Yet, neither in his memoir on "the combination of the matter of fire with evaporable fluids and the formation of aeriform elastic fluids" (*Mém. Acad. R. Sci.*, 1777, p. 420) nor in the memoir of 1780, to which we have referred above, did he make any reference to Black, although in the first of these he mentioned Richmann, Mairan and Cullen.

Robison towards the end of his criticisms added that Black announced his views only in his lectures and that, although urged to publish because his work was being plagiarized by others, he never went farther than preparing two octavo pages and some notes for continuing the account; and he wrote here of Wilcke's memoir of 1772, describing it on this occasion as referring to the melting of ice,[1] as if it were a mere appropriation of Black's discovery (pp. 523–4), whereas a knowledge of the contents would have informed him otherwise.

But the fine edge of Robison's sword was reserved for de Luc,[2] and his comments on this philosopher (pp. 524, 527–31), running to nearly five quarto pages of small type, provide an interesting example of misinformed and misdirected invective.[3] This, however, was taken so seriously by a reviewer of the *Lectures* that he asserted that the conduct imputed to de Luc by Robison "would have been deemed common imposture, if avarice, not vanity, had been the motive, and money, not fame, the end" (*Edinburgh Review*, 1803, **3**, 21). De Luc in a reply, which, contrary to the declared policy of his journal, the editor saw fit to publish and indeed to publish in full, cleared himself of all the charges preferred by Robison; showed that his only claim was to be the author of the meteorological theory that evaporation

---

[1] Cf. p. 36 above.

[2] J. A. de Luc (1729–1817). Born Geneva. Took up residence in London in 1770. Appointed reader to the Queen in 1773. Honorary Professor of Philosophy and Geology in Göttingen, 1798–1806. Returned to England and died at Windsor.

[3] The current edition of the *Encyclopædia Britannica* (article on Joseph Black) repeats this criticism, stating that Black did not publish any detailed account of his work on heat, "so that others, such as J. A. de Luc, were able to claim the credit of his results."

was produced by the combination of fire with water,[1] a theory that he had formulated in opposition to the notion that air acted as a solvent of moisture (*ibid.*, 1805, **6**, 502); and presented some interesting historical matter, showing that Laplace, Lavoisier, Monge and Vandermonde were well acquainted with Black's theory and experiments on latent heat in 1781 (previous to the publication of Lavoisier and Laplace's memoir) through the medium of Crawford's *Animal Heat* (1779), that it "was then in great agitation among these philosophers," that no doubts were expressed as to Black's priority, although Wilcke had published similar experiments, and that in French and German publications of 1800 and 1803 he (de Luc) had ascribed the discovery to Black.

It does not appear necessary to us to reproduce here the details of the controversy between Robison, Watt and de Luc; it is only proper to add that an examination of the literature completely clears de Luc of the charge of plagiarism, since his purely qualitative theory was formulated in entire independence of Black, to whom he repeatedly gave full credit for his work in this field.   But the curious in these matters may consult de Luc's *Recherches sur les Modifications de l'Atmosphère* (Geneva, 1772, I, pp. 350–1 and II, pp. 176, 179 and 182–3) and his *Idées sur la Météorologie* (London, 1786–7, I, 177–9 and 214–15), together with the *Appendice* (pp. 503–16) to Vol. I (not Vol. II, as Robison stated) of the latter work, noting that the pagination of the *Appendice* begins with p. 485, which follows p. 543 of the text.   A comment in a letter from Watt to de Luc is, however, for the light it sheds on the relations between Black and Watt, far too interesting to be passed over.   In this letter Watt referred to a passage in de Luc's *Météorologie* (I, p. 215), which stated that Black did no more than discover that the vapour of boiling water contained a great deal of heat and that he (Watt) had then carried out careful experiments to determine the amount of this heat.   Watt wrote that this was not the case; and he

---

[1] It was similarly suggested by Lemery the younger that water owed its fluidity to admixture with the "matter of fire" (*Mém. Acad. R. Sci.*, 1709, p. 400).

bluntly added that he had no merit in these researches other than that of having varied and frequently repeated these experiments to get an exact value for his own work, exactness not being at all necessary to Black's theory.

J'ai été d'autant plus précis sur ce sujet [reads the final part of the letter] qu'on n'a point apprécié les découvertes du Docteur comme elles le méritent, & que son extrême modestie a permis à d'autres, de donner comme leur appartenant, des Théories qu'ils avoient apprises le lui-même ou de ses Disciples. Or j'aurois peur que le passage de votre Ouvrage que je viens de rapporter, ne me rangeât dans ce nombre. C'est pourquoi je vous prie d'insérer dans votre Appendice la traduction de cette Lettre, comme étant un acte de justice envers le Docteur & envers moi (*Appendice*, p. 509).

Robison pointed out, too, that there was much written evidence to prove that Black's ideas preceded those of all the others whom he had named, and that the contents of certain note-books then in his possession and originally belonging to Black showed that the latter's doctrine of latent heat was not suddenly formed, but had occupied his attention continually during the years 1754 to 1757, when he was a student at Glasgow and at Edinburgh and during his first year as lecturer in Chemistry at Glasgow. Watt, said Robison, thought that Black had completed his proofs of the absorption and evolution of heat in liquefaction and congelation in 1758; but Robison found that in a sheet of Black's lecture-notes for that year the subject was still treated generally without any quantitative study. He was satisfied, however, that Black extended his views to the process of vaporization shortly after he had established the case in liquefaction; and, he added, "it is a fact that the doctrine of latent heat, precisely as it is contained in these lectures, was explained and demonstrated by Dr. Black in 1761 and in every subsequent course" (p. 527).

Casting about for evidence as to how these suspected plagiarists might have obtained their information, Robison stated that manuscript copies of Black's lectures were on sale for a moderate price, and suggested further and quite erroneously that both de Luc and Wilcke might have obtained their information from a surreptitious publication of Black's work and ideas that appeared in London in 1770. This book, barely mentioned by Robison, is of some interest and not without historical importance. It was published anonymously by Nourse in London in 1770, and, although it gave an incomplete account of Black's researches, it is the first printed version of them and precedes the publications of those who matter most here and, in particular, Wilcke. It bore the title *An Enquiry into the General Effects of Heat; with Observations on the Theories of Mixture*, which closely resembled that used by Black for his own discourses, namely, *Lectures on the Effects of Heat and Mixture*. The *Preface* described the theory of latent heat as "the invention of the ingenious Dr. Black" (p. iii); and the order of the sections of *Part I*, which dealt with heat, was identical with that followed by Black in his course of lectures. The anonymous author of this work appears to have compiled it with a copy or notes of Black's lectures before him, since whole phrases occurring in it are to be found in Robison's subsequent edition of the *Lectures*; he described Black's first quantitative experiments on the fusion of ice, giving the same value for the latent heat from slightly different figures (pp. 41–2), and also his first quantitative experiment on the vaporization of water, giving the value 790 (instead of 810), which he obtained by substituting 54° for 50° as the original temperature of the water, leaving the other figures as they stood (pp. 48–9). This volume, therefore, although anonymous, gave Black credit for his discoveries on latent heat, and since it was published in 1770, it preceded the authorized publications of all Black's contemporaries, imitators and others, in this field.

We have found a curious reference to this book in de Luc's *Idées sur la Météorologie* (I, *Appendice*, p. 511), in a passage forming part of a letter on the subject of Black's discoveries in heat from one of de Luc's friends who was connected with Black and well acquainted with his work. It reads:

Il parut chez Nourse, en 1770, sous le titre *Enquiry into the general Effects of Heat*, un extrait informe d'une partie de ses Leçons, contenant à-peu-près tout ce qu'il professe sur cette Théorie. Une Expérience qu'il attribue dans ses Leçons à M. Watt, y est rapportée à la première personne; ce qui avoit porté quelques lecteurs à penser, que M. Watt en étoit l'Auteur. Mais le Docteur ne s'y est point mépris; & notre ami ne connoissoit pas même la brochure quand je lui en ai parlé.

It might be noted here that Cavendish in certain researches [1] which he did not publish carried out work similar to that of Black (*Scientific Papers of the Honourable Henry Cavendish, F.R.S.*, Cambridge, 1921, II, pp. 326 ff.). These studies were either contemporary with Black's or possibly somewhat later, but it is reasonably certain, according to the editors of the *Scientific Papers, etc.*, that Cavendish's experimental work was completed in or before 1764. It appears that Cavendish carried out experiments on specific and latent heats by methods similar to Black's; and his biographer suggested that possibly "a reluctance to enter into even the appearance of rivalry with Black prevented him from publishing researches which might be thought to trespass upon ground which the latter had marked off for himself and pre-occupied" (G. Wilson, *Life of the Honble. Henry Cavendish*, London, 1851, p. 446). This suggestion appears all the more reasonable when we find that Cavendish referred to Black's observation that the water in the worm-tube of a still was heated more by the steam than by an equal quantity of boiling water, and stated that Black had "thence computed the quantity of heat generated by the condensation of steam," although he (Cavendish) was "not acquainted with the result of his experiment," adding "but I repeated it myself" (*Scientific Papers, etc.*, II, pp. 346–7, 358). But, since this was an observation fundamental to Black's theory and an observation made only in reference to that theory, and, indeed, made at this very time, the editorial statement in the *Scientific Papers* (II, p. 326) that Cavendish worked "to a large extent in ignorance of Black's observations and theories" is one that may quite fairly be doubted; and it seems reasonable to

[1] Attention was first drawn to this work by Vernon Harcourt in his Presidential Address to the B.A. Meeting (Birmingham), 1839 (*B.A. Report*, 1839, pp. 45–50); and a further account was given by G. Wilson (*Life of the Honble. Henry Cavendish*, London, 1851, pp. 446 ff.).

suppose that Cavendish had some knowledge of Black's work and theories at this time, i.e. previous to 1764, although no significance may be attached to any resemblance or any difference in the experimental methods of two such original workers.[1]

Enough has now been said to show Black's claims to the discovery of specific and latent heats; the next chapter, which deals with Wilcke's experiments on the melting of snow by hot water, will reveal the great superiority of Black's researches over those of his Swedish contemporary.

[1] Cavendish's value for the latent heat of vaporization of water was 982° (*Scientific Papers*, II, pp. 345–6) and for that of the condensation of steam was 920° (*ibid.*, pp. 346–7), both being "about 900°" (*ibid.*, p. 358), while the latent heat of fusion of snow was 170° (*ibid.*, p. 348), although in the latter case Cavendish's (later) published work indicates that he relied on another value which he had obtained, namely, 150° (*Phil. Trans.*, 1783, **73**, 312). Cavendish determined the latent heats of vaporization of water and condensation of steam by methods identical with those devised by Black, and the latent heat of fusion of snow by the method of mixture.

CHAPTER III

WILCKE'S DISCOVERY OF LATENT HEAT

We must now turn to the Baltic workers: Krafft and Richmann in St. Petersburg, Wilcke and Gadolin in Sweden. Of these, Krafft and Richmann were concerned solely with finding a formula for the temperature obtained on mixing different amounts of the same liquid (water) at different temperatures, and Wilcke, in the first instance, with modifying Richmann's formula to cover the case where one of the constituents of the mixture is water, the other snow. While our interest naturally centres on Wilcke, yet Wilcke's work is so closely linked to Richmann's and Richmann's to Krafft's, that both these earlier workers are entitled to a place in this account of a discovery of which neither ever dreamt.

In fact, it is no exaggeration to say that the first step towards Wilcke's discovery was taken some 110 years before the discovery was made; for Morin [1] in his *Astrologia Gallica* (The Hague, 1661) has a single chapter on the determination of the temperature of a mixture. Krafft came across this work, and was so struck by the extent to which Morin's result did actually represent the facts that he expressed it mathematically, and then chose an expression of the same general form as the starting-point for his own investigation. Morin's work is, indeed, of uncommon interest; for we have in it the earliest attempt to

---

[1] Jean Baptiste Morin, M.D. (1583–1656), Professor of Mathematics and Astronomy at the Collège Royal, Paris, founder of the science of longitudes and editor, after Kepler, of the Rudolphine Tables. In addition to the *Astrologia Gallica*, which took thirty years to complete, Morin wrote several works on astronomy.

deal mathematically with a quantitative problem in heat. To quote Morin's own words: *Hac de re (quod sciam) nemo hactenus quidquam determinavit.*[1] As the passage in question is of no great length, we have included the full text as an Appendix.[2]

In the present chapter we propose to discuss in turn the work of Morin, Krafft, Richmann and Wilcke on Mixtures. Wilcke's work on Specific Heat, with that of his contemporary, Gadolin, will form the subject of a separate chapter.

Jean Morin (1661). It may be said at once that Morin's argument is by no means easy to follow. This is largely due to the very confusing way in which any temperature is referred to as being "so many degrees of heat and so many degrees of cold." Lambert, in discussing the thermometers of this period, says that they were divided into eight degrees, the middle division being "temperate"; that there were four "degrees of cold" (chilly, cold, very cold, intense cold) below "temperate," and four "degrees of heat" (mild, warm, very warm, hot) above "temperate" (*Pyrometrie*, Berlin, 1779, p. 16). This statement is not however supported by Morin, who says, indeed, that philosophers were generally agreed that a mixture could not have more than eight degrees of the "contrary qualities," but who attaches an entirely different meaning to a "degree of heat" or "degree of cold." To Morin, a temperature corresponding to the mid-point of the thermometer would be one of "four degrees of heat and four degrees of cold," "four degrees of heat" because it is four degrees above the lowest point, and "four degrees of cold" because it is four degrees below the highest point. Similarly, a mixture whose temperature corresponded to the sixth division (i.e. "warm") would have six degrees of heat and two degrees of cold. When Morin says that a mixture cannot have more than 8 degrees of the contrary qualities, he means that a mixture cannot have, for instance, 5 degrees of heat and 4 degrees of cold.

[1] No one, as far as I know, has hitherto determined anything on this matter.

[2] See pp. 149–51 below.

Morin wishes to determine the temperature that would result on mixing cold water containing 2 degrees of heat and 6 degrees of cold (i.e. corresponding to the second division on the thermometer) with an equal quantity of warm water containing 4 degrees of heat and 4 degrees of cold (i.e. corresponding to the mid-point on the thermometer).

He argues that the mixture could not have 2 degrees of heat (i.e. remain at the temperature of the colder water), for then 2 of the degrees of heat in the hot at 4 (degrees of heat) would have been destroyed, the 4 degrees of heat in the hot having had no effect on the 6 degrees of cold in the cold.   Nor could the mixture have 4 degrees of heat (i.e. remain at the temperature of the warm water), for then the cold would have had no effect. The mixture must therefore have more than 2 degrees of heat, but less than 4 degrees of heat.

The argument by which Morin rejects the correct value of 3 degrees of heat is interesting.   It amounts to this: if the mixture had 3 degrees of heat (and therefore 5 degrees of cold), the 6 degrees of cold in the cold water would have had the effect of reducing the 4 degrees of heat in the warm water by 1 degree, while the 4 degrees of heat in the warm water would have had the effect of reducing the 6 degrees of cold in the cold water by 1 degree.   Six degrees of cold and 4 degrees of heat would therefore have produced effects of equal intensity in the same interval of time.   This Morin holds to be impossible, as 6 degrees of cold is stronger than 4 degrees of heat; also the difference between the cold in the two constituents is relatively less (one is $\frac{2}{3}$ of the other) than the difference between the heat (one is $\frac{1}{2}$ the other) so that on mixing, when cold becomes equal to cold and heat to heat, the change in the cold will not be as great as the change in the heat, that is, the 6 degrees of cold will be less affected by the 4 degrees of heat than the 4 degrees of heat by the 6 degrees of cold.   The resulting temperature should therefore be somewhat lower than 3 degrees.

Morin bases his solution of the problem on three principles, which are:

(1) No destruction of the qualities (heat and cold) occurs in the act of mixing, only an interchange.

(2) Action and reaction take place only between the opposite qualities present in greater degree; those present in lesser degree become intensified.

(3) The total "virtue" of any number of degrees of heat is equal to the total "virtue" of the same number of degrees of cold.

Applying these principles, Morin proceeds as follows:

Action and reaction take place only between the 6 degrees of cold and the 4 degrees of heat (by 2).

The total virtue of 6 degrees of cold is $1\frac{1}{2}$ times that of 4 degrees of heat (by 3).

Two numbers are therefore required, which shall be in the ratio 3 : 2.

Let them be 3A and 2A.

Then 3A represents the cold that the 6 degrees of cold in the cold water produce in the 4 degrees of heat in the warm water. This, added to the 4 degrees of cold already existing in the warm water, gives $3A + 4$ for the total cold in the warm water.

2A represents the heat that the 4 degrees of heat in the warm water produce in the 6 degrees of cold in the cold water. This, subtracted from the 6 degrees of cold, gives $6 - 2A$ for the total cold of the cold water.

But when the waters are mixed, the cold in the one is equal to the cold in the other.

Hence $3A + 4$ is equal to $6 - 2A$.

The solution of this equation is worth quoting in full:

"Give 2A to each: then you will have $5A + 4$ equal 6. Take 4 from each: then you will have 5A equal 2. Then divide 2 by 5 and you will have $\frac{2}{5}$ for the value of 1A. Wherefore the value of 3A will be $\frac{6}{5}$, and the value of 2A will be $\frac{4}{5}$, and these are the required numbers."

On substituting in either of the expressions in the equation, the value $5\frac{1}{5}$ is obtained for the cold in the mixture. The heat is therefore $2\frac{4}{5}$ degrees.

Morin then goes on to consider briefly the case where the hot and cold water are not equal in quantity. If the cold at 6 (degrees of cold) is double the hot at 4 (degrees of heat), then the total virtue of the cold must be doubled. Thus, if cold water be mixed with $1\frac{1}{2}$ times the quantity of hot water, the hot being at 6 (degrees of heat) and the cold at 7 (degrees of cold), then the mixture will have $3\frac{13}{16}$ degrees of heat and $4\frac{3}{16}$ degrees of cold. Morin does not give the actual calculation, but it should doubtless be as follows:

The total virtue of $1\frac{1}{2}$ measures of hot water at 6 (degrees of heat) : total virtue of 1 measure of cold at 7 (degrees of cold)

$$= 6 \times 1\frac{1}{2} : 7$$
$$= 9 : 7$$

Hence we require two numbers in the ratio 9 : 7.

Let them be 9A and 7A.

Then 9A is the heat given by the hot to the cold.

Therefore the heat content of the cold is 9A + 1 (for the cold already contains 1 degree of heat).

And 7A is the cold given by the cold to the hot.

Therefore the heat content of the hot is reduced to 6–7A.

But, after mixing, the heat content of the one is equal to the heat content of the other.

Therefore 9A + 1 = 6–7A.

Whence $\qquad$ A $= \frac{5}{16}$,

so that the mixture has $\frac{5}{16} \times 9 + 1$ or $3\frac{13}{16}$ degrees of heat, and therefore $4\frac{3}{16}$ degrees of cold, as stated by Morin.

If we take one of Morin's degrees as equal to $12\frac{1}{2}$ degrees Centigrade, the above result becomes $48°$ C. Calculation by the usual method of heat exchange gives $50°$ C., so that Morin's result is remarkably near. We have carried out a number of calculations on the lines of the examples given, and find that the temperatures obtained are rather too low when below the mid-point, too high when above it, and correct at the mid-point.

It is rather surprising that Morin's method should give such good results; for the fallacy involved in his third principle is serious. This principle states that the total virtue of any number of degrees of heat is equal to the total virtue of the same number of degrees of cold. Applying this to his first example (pp. 56–7 above), he says that the total virtue of 6 degrees of cold is $1\frac{1}{2}$ times the total virtue of 4 degrees of heat, and he therefore seeks two numbers in the ratio 3 : 2. His solution is that the presence in the mixture of the cold water with its 6 "degrees of cold" decreases the heat of the hot water by $\frac{6}{5}$ of a degree, while the presence of the equal quantity of hot water with its 4 "degrees of heat" decreases the cold of the cold water by $\frac{4}{5}$ of a degree. The position becomes clearer if we remember that 6 degrees of cold are accompanied by 2 degrees of heat (that is, the *temperature* is 2 degrees) and that to decrease the cold means to increase the accompanying heat. In effect, Morin says that water at 2 degrees (on his scale) mixed with an equal quantity of water at 4 degrees gives a temperature of $2\frac{4}{5}$ degrees. He rejects the correct mean value of 3 degrees on the grounds that it violates this same principle. It is interesting to apply Morin's reasoning to an example on the Centigrade scale. Suppose water at 50° C. is mixed with an equal quantity at 70° C. Then, as the cold water is 50 degrees *below the upper fixed point* and the hot water is 70 degrees *above the lower fixed point*, the temperature of the mixture should, according to him, be $\frac{5}{12}$ of $(70 - 50)$ or $8\frac{1}{3}$ degrees below 70° C. and $\frac{7}{12}$ of $(70 - 50)$ or $11\frac{2}{3}$ degrees above 50° C., both of which give a temperature of $61\frac{2}{3}$° C., surprisingly close to the correct value of 60° C.

We can readily sympathize with Krafft's amazement at finding anything so sound "in the midst of so much astrological nonsense," though we would spare contempt for the beliefs of another age.

Georg Wolffgang Krafft (1744). After a few introductory remarks on Morin, Krafft in his *De Calore ac Frigore Experimenta Varia (Comment. Acad. Sci. Imp. Petrop.*, 1744–6, **14**, 218) goes on to show that Morin's solution of the problem of finding the temperature θ that results from mixing two unequal quantities of water m_1, m_2 at temperatures t_1, t_2 respectively (on Morin's scale), where m_1, t_1 refer to the colder water, m_2, t_2 to the hotter, can be expressed by the formula

$$\theta = \frac{m_2 t_2{}^2 + \overline{8 - t_1}.m_1 t_1{}^1}{m_2 t_2 + \overline{8 - t_1}.m_1}.$$

[1] In Krafft's symbols:

$$\text{Temperature of mixture} = \frac{n^2 b + \overline{8 - m}.ma}{nb + \overline{8 - m}.a},$$

where a, b denote the number of measures of cold and hot water respectively, and m, n the corresponding temperatures or degrees of heat, so that $8 - m$ denotes the number of degrees of cold in the cold water.

After some hesitation we have abandoned the symbols used by Krafft, Richmann and Wilcke in favour of those in general use in England at the present time. This departure from strict fidelity to the originals on quite a minor point seems justified by the increased ease with which the argument can be followed.

When the various gaps have been filled in, Krafft's deduction of the given expression is as follows:

On mixing, let x be the cold produced by the cold water in m_2 measures of hot water at t_2 degrees of heat (Morin's 3A); and let y be the heat produced by the hot water in m_1 measures of cold water at t_1 degrees of heat, that is, at $(8 - t_1)$ degrees of cold (Morin's 2A).

Then, by Morin's argument:

$$x : y = (8 - t_1).m_1 : m_2 t_2,$$

whence

$$y = \frac{m_2 t_2 x}{(8 - t_1)m_1} \qquad \cdots \cdots \cdots \quad (1)$$

t_1 is the number of degrees of heat in the cold water, and t_2 is the number of degrees of heat in the hot water.

Hence, on mixing:

The heat content of the cold water is $y + t_1$

and the heat content of the hot water is $t_2 - x$.

These are equal.

Therefore

$$y + t_1 = t_2 - x,$$

whence

$$y = t_2 - t_1 - x, \qquad \cdots \cdots \cdots \quad (2);$$

and on eliminating y between (1) and (2), we have

$$x = \frac{(8 - t_1)(m_1 t_2 - m_1 t_1)}{m_2 t_2 + 8 m_1 - m_1 t_1} \qquad \cdots \cdots \quad (3)$$

The "cold" in the hot water at t_2 is $8 - t_2$.

Therefore the cold in the mixture is $x + (8 - t_2)$.

Therefore the heat in the mixture (that is, its temperature) is

$$8 - (x + 8 - t_2);$$

and on substituting in this the value for x given by (3), we obtain Krafft's expression.

Krafft wished to obtain a formula for the temperature of the mixture in a more scientific manner, and one adapted to Fahrenheit's thermometer. He assumed the expression to be of the general form

$$\theta = \frac{\alpha t_1 + \beta t_2}{\gamma m_1 + \delta m_2}, \quad \cdots \quad \text{(A)}$$

where θ is the temperature of the mixture in degrees Fahrenheit, m_1, m_2 the number of measures of cold and hot water respectively, t_1, t_2 the corresponding temperatures in degrees Fahrenheit, and α, β, γ, δ are coefficients whose values remain to be determined. He then proceeded as follows :

I. If the two measures are at the same temperature (i.e. if $t_2 = t_1$), then the temperature of the mixture will also be t_1.

Hence in this case we have $\theta = t_1$, and (A) therefore becomes

$$t_1 = \frac{\alpha t_1 + \beta t_1}{\gamma m_1 + \delta m_2},$$

whence $\beta = \gamma m_1 + \delta m_2 - \alpha$

and $\alpha = \gamma m_1 + \delta m_2 - \beta$.

On substituting the value for β in (A) we obtain

$$\theta = \frac{(\gamma m_1 + \delta m_2)t_2 - (t_2 - t_1)\alpha}{\gamma m_1 + \delta m_2}, \quad \cdots \quad \text{(B)}$$

and on substituting the value for α in (A) we obtain

$$\theta = \frac{(\gamma m_1 + \delta m_2)t_1 + (t_2 - t_1)\beta}{\gamma m_1 + \delta m_2} \quad \cdots \quad \text{(B}^1\text{)}$$

II. If the quantity of the hot water added to the mixture is infinitesimally small in comparison with the quantity of the cold water, then the temperature of the mixture will be the temperature of the cold water (t_1). In this case m_2 is negligible, so that (B) becomes

$$t_1 = \frac{\gamma m_1 t_2 - (t_2 - t_1)\alpha}{\gamma m_1},$$

whence $\gamma m_1 = \alpha$.

III. If the quantity of the cold water added to the mixture is infinitesimally small in comparison with the quantity of the hot water, then the temperature of the mixture will be the temperature of the hot water (t_2). In this case m_1 is negligible, so that (B^1) becomes

$$t_2 = \frac{\delta m_2 t_1 + (t_2 - t_1)\beta}{\delta m_2},$$

whence $\delta m_2 = \beta$.

IV. On substituting $\alpha = \gamma m_1$, and $\beta = \delta m_2$ in (A), we obtain

$$\theta = \frac{\gamma m_1 t_1 + \delta m_2 t_2}{\gamma m_1 + \delta m_2} \quad . \quad . \quad . \quad . \quad \text{(C)}$$

The interesting part of Krafft's work lies in his determination of the remaining coefficients γ and δ. For this purpose he appealed to experiment.

Two equal quantities of water, one at $44°$ F. and the other at $120°$ F., were mixed, and the temperature of the mixture was found to be $76°$ F.

On putting $m_1 = m_2$,
$$t_1 = 44,$$
$$t_2 = 120,$$
and $\qquad \theta = 76$

in (C), it is found that $\dfrac{\gamma}{\delta} = \dfrac{11}{8}$,

and this, on substitution in (C), gives

$$\theta = \frac{11 m_1 t_1 + 8 m_2 t_2}{11 m_1 + 8 m_2},$$

which is Krafft's formula.[1]

Krafft then carried out a number of experiments to test the accuracy of his formula. It will be sufficient to give one of these here.

Two equal quantities of water, at temperatures of $64°$ F. and $111\frac{1}{2}°$ F., gave on mixing a temperature of $84°$ F., which is

[1] In Krafft's symbols:

$$\text{Temperature of mixture} = \frac{11am + 8bn}{11a + 8b}.$$

PLATE III.

GEORG WOLFFGANG KRAFFT (1701–54)
Professor of Mathematics and later of Theoretical and Experimental Physics at St. Petersburg. Subsequently Professor of Mathematics at Tübingen.

(Reproduced by kind permission of the Academy of Sciences, Leningrad.)

also the value found by calculation from the formula. The agreement was equally good when unequal quantities of water were used.

Now, Krafft gives no experimental details whatever: he merely states the quantities and temperatures of the water taken and the temperature that resulted on mixing. It therefore occurred to us that we should gain some insight into his experimental methods by carrying out the experiments with increasing care, especially in view of the discrepancies between Krafft's results and those obtained by Richmann. Like Krafft, we took one or more measures (a measure being in our case 50 c.c.) of boiling water, poured it into several measures of colder water at a known temperature, and noted the maximum temperature registered by the thermometer. We began by carrying out the experiments as crudely as possible, merely taking care to pour the hot water as soon as it boiled into the cold water, contained in a glass beaker standing on a wooden bench. We had no occasion to go beyond this stage, for we found in every case a most pleasing agreement between the temperatures observed and those calculated from Krafft's formula. For example, 1 measure of water at 212° F., added to 3 measures of water at 123·8° F., gave an observed temperature of 140·9° F., the calculated temperature being 141° F.

Krafft appears to us, indeed, as essentially a "practical man," who avoids subtleties in his arguments as he avoids needless refinements in his experiments, would regard them, perhaps, as "speculative nonsense." To him the thermometer is everything: a little hot water in a large quantity of cold will not affect the temperature to any appreciable extent. His formula is a practical man's good working rule: if you pour some boiling water into some cold water, you want to know what the temperature is going to be, not what it might be under ideal circumstances. Any increase in the refinement of the experiments would be bound to give temperatures higher than those predicted by Krafft's formula, as Richmann found.

Temperature and " Quantity of Heat." With Richmann, the distinction between temperature and quantity of heat becomes for the first time important. To Morin, heat and cold are "opposite qualities" or "contraries"; they belong to the "similars and dissimilars" so dear to the philosophers of the century before; they cannot be destroyed, but they can blend together, as in a mixture, when the predominating "opposites" dilute each other, while the lesser become enhanced; their intensities can be measured with a thermometer, and, of a pair of "opposites," it is the more intense that predominates on blending. To Morin there could be no question of *quantity* of heat or cold, though something of the idea seems to be creeping in when he argues that twice the quantity of water would have twice the total virtue—but even here it is rather that twice the quantity of water will produce twice the effect. Perhaps we can appreciate the point of view of Morin's age if we consider *colour* instead of heat. Colour to us is a quality; we can imagine a scale in which blue, for example, is graduated in depth from very pale to very dark; by comparison with our scale we could define any given sample of blue as having so many "degrees of blue"; we could say that 2 degrees of blue mixed with 4 degrees of blue would give a blue whose intensity was about 3; if we think of dyes, we could also argue that 2 measures of blue dye at 6 degrees would have twice the bluing effect of 1 measure—but it would surely never occur to us to try to estimate the *quantity of blueness* in a thing.

Roughly midway between Morin, to whom quantity of heat could have no more meaning than quantity of colour has to us, and Wilcke, to whom quantity of heat, as distinct from temperature, was a very real thing, capable of measurement as soon as a suitable standard of measurement had been found, stands Richmann. How far these two notions, intensity and quantity, were separated in Richmann's mind, is not easy to decide. To a memoir dealing purely with the temperature of a mixture he gives the title "On the *Quantity* of Heat" (*De Quantitate Caloris*); he assumes that, if the same degree of "heat" (*idem caloris gradus*) is uniformly distributed through 2, 3, 4, etc. times as much water, the "heat" produced (*calor hinc generatus*) will be $\frac{1}{2}, \frac{1}{3}, \frac{1}{4}$, etc. what it would be in unit quantity of water.

Here *caloris gradus* must be taken to mean *quantity of heat*, and *calor hinc generatus* to mean temperature. Evidently Richmann's contemporaries were not likely to be confused by the use of a single word to cover what, to us, are two very different things. And this is understandable; for in his day there was only one thing about heat to which a definite, determinable value could be attached, and that was temperature. There was plenty of speculation as to the nature of heat about this time, some holding it to be a substance, others some form of motion; those who held the former view might, for convenience in argument, refer to "quantity of heat," but they could assign no measurable value to it; while to those who held the latter view, "quantity" would be almost as meaningless as it would have been to Morin. To Richmann, the only "quantity" about heat is temperature; a memoir dealing with temperature is quite fittingly described as *De Quantitate Caloris*, and in discussing the matter, the units (*gradus*) in which the quantity is measured may be inserted or not, according to fancy. Richmann is seeking a mathematical expression for the temperature of a mixture in terms of the masses and temperatures of the constituents; it is useless for him to introduce into his argument any term representing quantity of heat in our sense of the phrase, since this would appear in his equation and could be eliminated only by expressing it in terms of his measurable quantities, mass and temperature—a thing that he could not do. In our account of Richmann's work we have used sometimes *temperature*, sometimes *heat*, whichever seems the more appropriate; but it should be borne in mind that Richmann himself uses the term *calor* to cover both.

G. W. Richmann (1744–1747). Richmann's deduction of his formula for the temperature of a mixture was published in his *De Quantitate Caloris, quae post miscelam fluidorum, certo gradu calidorum, oriri debet, cogitationes* (*Nov. Comment. Acad. Sci. Imp. Petrop.*, 1747–8, **1**, 152). Actually, Richmann had arrived at his formula before Krafft's was published, but he had laid his notes aside and forgotten them until Krafft's memoir recalled them to his mind. When he tried Krafft's figures in his own formula, he found that the calculated temperatures were always too high; and this led him to take up the work again. Though this memoir did not appear until 1747, Richmann says in another memoir (*ibid.*, p. 168) that his work on Mixtures was read to the Academy on October 19th, 1744.

Richmann begins by assuming that the heat is distributed uniformly throughout the liquid. If, therefore, the same heat (*idem caloris gradus*) is distributed through a mass 2, 3, 4, etc. times as great, the temperature (*calor hinc generatus*) will be reduced to $\frac{1}{2}$, $\frac{1}{3}$, $\frac{1}{4}$, etc.—that is, the temperature (*calor*) is inversely proportional to the mass.

Let the mass of one portion of liquid be m_1.

Let its temperature (*calor distributus per hanc massam*) be t_1.

Then if the same "heat" were distributed through a mass $m_1 + m_2$, the temperature (*calor hinc generatus*) would be $\dfrac{m_1 t_1}{m_1 + m_2}$.

Let the mass of a second portion be m_2, and let its temperature be t_2.

Then this "heat," distributed through a mass $m_1 + m_2$, would give rise to a temperature $\dfrac{m_2 t_2}{m_1 + m_2}$ (*erit calor, a calore n per massam a + b distributo, ortus* $= \dfrac{bn}{a + b}$).[1]

If the "heat" t_1 of the mass m_1 and the "heat" t_2 of the mass m_2 are together uniformly distributed through the mass $m_1 + m_2$, then the sum of the "heats" t_1 and t_2 in the mixture ought to be equal to $\dfrac{m_1 t_1 + m_2 t_2}{m_1 + m_2}$ (*calor in hac massa, sive in mixto a et b, aequalis esse debet summae calorum m + n distributorum per massam a + b, sive* $= \dfrac{ma + nb}{a + b}$).

----

[1] We have kept to Richmann's symbols in the passages quoted. He follows Krafft in using *a, b* for the masses, *m, n* for the corresponding temperatures.

PLATE IV.

Frid. Guil. Richmann Prof. de la Phys. Experim.
dans l'acad. Impl. des Scienc. à P.de rbourg.
tué par un coup de foudre 1753. le 3. Juillet.
Jupiter qui tonne, escampota la poudre,
Lui poudre son eclair et lui pompa la foudre
Souffla

GEORG WILHELM RICHMANN (1711–53)

Studied at Halle and Jena. Professor of Experimental Philosophy in the Imperial Academy of St. Petersburg (1745–53). Killed on August 6, 1753, during a severe thunderstorm at St. Petersburg by a discharge from the apparatus that he had devised for the collection and measurement of atmospheric electricity, while exhibiting this apparatus in working order to the engraver, Sokolov, who was preparing the figures for a projected work on this subject. The peculiar circumstances of Richmann's death were the subject of two papers in the *Phil. Trans.* (1754, **48**, 765 and 1755, **49**, 61).

(Reproduced by kind permission of the Academy of Sciences, Leningrad.)

Richmann's Formula for the temperature of a mixture is therefore

$$\theta = \frac{m_1 t_1 + m_2 t_2}{m_1 + m_2},$$

where m_1, m_2 are the masses,

t_1, t_2 the corresponding temperatures,

and θ is the temperature of the mixture.

There are several points in Richmann's argument that present no little difficulty. Even the significance of his assumption is not immediately apparent. If we take it to mean that the liquid has the same temperature at all points, it is trivial and leads nowhere. If, on the other hand, we take it to mean that unit mass of the liquid contains everywhere the same quantity of heat, it still seems pointless, unless we make the additional assumption that temperature is proportional to the quantity of heat per unit mass, when it would follow that, for a given quantity of heat, the temperature is inversely proportional to the mass. This may be what Richmann had in mind. A consideration of the mathematical expressions involved seems however to suggest something rather different. If we take the expression $\dfrac{m_1 t_1}{m_1 + m_2}$, we know definitely that t_1 is the temperature in degrees Fahrenheit of the water whose mass is m_1. In fact, t_1 is the only "heat" (*calor*) of which Richmann knows anything, or to which he can assign a numerical value. Then the "heat" t_1, multiplied by the number m_1, which is the number of units of mass in the water, might very well be taken to represent the total "heat" in the water—a quantity which need not be defined precisely, and to which a definite numerical value cannot be assigned, as it depends on an arbitrarily chosen unit of mass, and is, indeed, equal to the temperature of the water if the mass of the water is taken as unity. This total "heat" is not *quantity of heat*, but is a multiple of the temperature; or, equally well, temperature (measurable "heat") is a submultiple of the total "heat" (indeterminate), the submultiplier being the number of units of mass in the liquid—which fits in with the assumption that " 'heat' is uniformly distributed through the mass."

Again, if the total "heat" $m_1 t_1$ is now uniformly distributed through the mass $m_1 + m_2$, the "heat" that results has a determinate numerical value, since it is independent of the unit of mass chosen; but it cannot be measured thermometrically, since in practice any addition to the mass of the liquid would involve the addition of the indeterminate amount of "heat" in the water added. But if the added water has "heat" $\dfrac{m_2 t_2}{m_1 + m_2}$, the sum of the two calculable, though in practice unmeasurable, "heats" is again measurable thermometrically and is, in fact, the temperature of the mixture.

Having found a formula for the temperature of a mixture consisting of two quantities of water at different temperatures, Richmann extends it to include mixtures of any number of such quantities, the formula then becoming:

$$\text{Temperature} = \frac{m_1 t_1 + m_2 t_2 + m_3 t_3 + \text{etc.}}{m_1 + m_2 + m_3 + \text{etc.}},$$

where m_1, m_2, m_3 etc. are the masses.
and t_1, t_2, t_3 etc. the temperatures.

Richmann then points out that, in actual experiments, the following must be taken into account:

(1) Some of the heat of the mixture is taken up by the vessel and the thermometer.

(2) The heat of the vessel and thermometer is distributed among the mixture, vessel and thermometer.

(3) Some of the heat passes into the air during the time taken over the experiment. The loss of heat due to this cause can be neglected if the time taken is short.

When the masses of liquid involved are small, neglect of (1) may lead to serious discrepancies between the calculated and observed temperatures. Allowance must therefore be made for the vessel and thermometer, and Richmann proceeds as follows:

Let m_1, t_1 be the mass and temperature of one quantity of the liquid,

$\quad m_2$, t_2 be the mass and temperature of the other quantity of the liquid,

$\quad\quad m_t$, t_t ,, ,, ,, ,, ,, ,, thermometer,
and $\quad m_v$, t_v ,, ,, ,, ,, ,, ,, vessel.

Then the temperature that ought to be shown by the thermometer is

$$\frac{m_1 t_1 + m_2 t_2 + m_t t_t + m_v t_v}{m_1 + m_2 + m_t + m_v},$$

and the true temperature of the mixture, had no loss occurred, ought to be greater than this by an amount

$$\frac{m_1 t_1 + m_2 t_2}{m_1 + m_2} - \frac{m_1 t_1 + m_2 t_2 + m_t t_t + m_v t_v}{m_1 + m_2 + m_t + m_v}.$$

This, in Richmann's opinion, is the reason why Krafft's formula always gives temperatures lower than those given by his own formula; for Krafft's allows automatically for the loss of heat to the vessel and thermometer, whereas his (Richmann's) gives the theoretical temperature and requires a correcting term for the loss of heat that occurs in practice. In fact, by taking Krafft's vessel to have been equivalent to one measure of water and his thermometer to half a measure of water, Richmann showed that his own formula gave in many instances the same temperature for the mixture as did Krafft's, whose calculated values were in fairly close agreement with his observed temperatures. Thus, to take one or two examples from Richmann's Tables:

Temperature of mixture observed by Krafft	68° F.
Temperature calculated from Krafft's formula	68·12° F.
Temperature calculated from Richmann's formula	76·00° F.
Temperature calculated from Richmann's formula when allowance was made for vessel and thermometer	68·15° F.

Again:

Temperature of mixture observed by Krafft	80° F.
Temperature calculated from Krafft's formula	79·42° F.
Temperature calculated from Richmann's formula	82·00° F.
Temperature calculated from Richmann's formula when allowance was made for vessel and thermometer	81·03° F.

And lastly:

Temperature of mixture observed by Krafft	129° F.
Temperature calculated from Krafft's formula	130·54° F.
Temperature calculated from Richman's formula	131·27° F.
Temperature calculated from Richmann's formula when allowance was made for vessel and thermometer	131·00° F.

The last example is one of several in which the volume of the mixture was large, amounting nearly to 90 cubic inches. The small difference in the temperatures calculated from Richmann's two formulæ is due, as Richmann points out, to the comparatively small part played by the vessel when a large quantity of water is taken.

The only experimental results referred to by Richmann in the memoir that we have just been discussing are those obtained by Krafft. Richmann's own experiments on mixtures were carried out later and were published in his next memoir: *Formulae pro gradu excessus caloris supra gradum caloris mixti ex nive et sale amoniaco post miscelam duarum massarum aquearum diverso gradu calidarum confirmatio per experimenta (Nov. Comment. Acad. Sci. Imp. Petrop., 1747–8, 1, 168).* He begins this with the statement that, in confirming the formula that he announced to the Academy on October 19th, 1744, he had made use of Krafft's experimental results, and that he is now about to communicate to the Academy the experiments that he himself made later on for this purpose. In the last paragraph of the memoir there is a note on the two thermometers used in the experiments. Richmann says that these were very sensitive mercury thermometers, graduated according to Fahrenheit's scale and excellently constructed by Prins, the instrument-maker of Amsterdam. The bulbs were cone-shaped, the greatest thickness being about 4 London lines [1] and the length 12 lines: their volume was not greater than that occupied by 30 grains of water.

Richmann's account of his experiments is very brief. Six experiments only are given, and of these the first two are concerned solely with the rate of cooling of water in air. They are as follows:

Experiment I

An earthenware vessel was filled with water to a depth of 4 London inches. The water weighed 12 ounces. The diameter of the exposed circular surface was about 3 inches. The initial temperature was 128° F. and the temperature of the

[1] 1 line = $\frac{1}{12}$ inch.

air 66° F.   It took four hours for the water to cool to 67° F., thus:

In   5 minutes it cooled from 128° to 122°
,,   10   ,,   ,,   ,,   ,,   ,, 116°
,,   15   ,,   ,,   ,,   ,,   ,, 110°
,,   20   ,,   ,,   ,,   ,,   ,, 108°
,,   25   ,,   ,,   ,,   ,,   ,, 104°
,,   30   ,,   ,,   ,,   ,,   ,, 101°
,,   35   ,,   ,,   ,,   ,,   ,, 98°
,,   40   ,,   ,,   ,,   ,,   ,, 96°
,,   45   ,,   ,,   ,,   ,,   ,, 94°
,,   50   ,,   ,,   ,,   ,,   ,, 92°
,,   55   ,,   ,,   ,,   ,,   ,, 90°
,,   60   ,,   ,,   ,,   ,,   ,, 88°
,,   65   ,,   ,,   ,,   ,,   ,, 86°
,,   70   ,,   ,,   ,,   ,,   ,, 84½°
,,   90   ,,   ,,   ,,   ,,   ,, 83°
,,   110   ,,   ,,   ,,   ,,   ,, 82°
,,   135   ,,   ,,   ,,   ,,   ,, 80°
,,   165   ,,   ,,   ,,   ,,   ,, 79°
,,   200   ,,   ,,   ,,   ,,   ,, 78°, etc.

Experiment II

The same earthenware vessel was filled to the same depth as before with water at 116° F.   The temperature of the air was 67° F.   The times for equal decrements in temperature were observed, thus:

The temperature fell from 116° to 112° in   3 minutes
,,   ,,   ,,   ,,   ,,   ,, 108° ,,   7   ,,
,,   ,,   ,,   ,,   ,,   ,, 104° ,, 12   ,,
,,   ,,   ,,   ,,   ,,   ,, 100° ,, 18½   ,,
,,   ,,   ,,   ,,   ,,   ,, 96° ,, 26¼   ,,
,,   ,,   ,,   ,,   ,,   ,, 92° ,, 35   ,,
,,   ,,   ,,   ,,   ,,   ,, 88° ,, 45   ,,
,,   ,,   ,,   ,,   ,,   ,, 84° ,, 56   ,,
,,   ,,   ,,   ,,   ,,   ,, 80° ,, 72   ,,
,,   ,,   ,,   ,,   ,,   ,, 76° ,, 90   ,,

In both these experiments [says Richmann] we perceive that the temperature falls off more quickly the greater the difference between the temperature of the water and that of the surrounding air, or the temperature decreases more rapidly in the earlier intervals of time, more slowly in the later. But it does not seem possible to determine from the experiments in what ratio the decrements stand to equal intervals of time and in what ratio the intervals of time stand to equal decrements, whether in a constant ratio or not (*loc. cit.*, p. 170).[1]

Experiment III

Twenty-four ounces of water at 64° F. were poured into 12 ounces at 178° F. The temperature of the mixture was found to be 100° F. The temperature of the air was 66° F.

Experiment IV

Equal masses of water at 88° F. and 172° F. were mixed. The temperature of the mixture was found to be 126° F. The temperature of the air was 66° F.

Experiment V

Water at 77° F. was mixed with twice its weight of water at 156° F. The temperature of the mixture was found to be 126° F. The temperature of the air was as before.

Experiment VI

Water at 148° F. was mixed with twice its weight of water at 70° F. The temperature of the mixture was found to be 94° F. The temperature of the air was as before.

Richmann then compares the results obtained in experiments III, IV, V, and VI with those given by the various formulæ, namely, his own, $\left(\dfrac{m_1 t_1 + m_2 t_2}{m_1 + m_2} \right)$, Krafft's $\left(\dfrac{11 m_1 t_1 + 8 m_2 t_2}{11 m_1 + 8 m_2} \right)$

---

[1] Richmann did not let the matter rest here. His next memoir (*Nov. Comment. Acad. Sci. Imp. Petrop.*, 1747–8, **1**, 174) describes the many experiments that he carried out later in order to determine the Law of Cooling. Although this work constitutes Richman's most elaborate contribution to the science of heat, it is impossible to discuss it here, as to do so at all adequately would require more space than its slight bearing on our subject would justify. The Law at which Richmann arrived has already been quoted (see p. 44 above, footnote).

and that derived from Boerhaave's Rule [1] $\left(\dfrac{t_2 - t_1}{2}\right)$, where, in each case,

m_1, m_2 are the masses, and

t_1, t_2 the corresponding temperatures in degrees F.

Thus:

Experiment.		Richmann.	Krafft.	Boerhaave.
III	100	102	$94\frac{2}{5}$	—
IV	126	130	$123\frac{7}{19}$	42
V	126	$129\frac{2}{3}$	$123\frac{22}{27}$	—
VI	94	96	$90\frac{4}{25}$	—

Only one value is given under Boerhaave, as Experiment **IV** is the only experiment in which equal quantities were taken (the only case to which Boerhaave's Rule is applicable). Richmann points out that this single result is sufficient to show that Boerhaave is entirely wrong, as the temperature of the mixture calculated by his rule is lower than that of the cooler water (88°).

These experiments, Richmann considers, sufficiently demonstrate the superiority of his own formula over Krafft's. For his own values are always slightly higher than those found in the experiments, which is to be expected, as the vessel absorbs some of the heat. The fact that the values calculated by Krafft's formula are *lower* than the observed values is in itself sufficient to prove the inaccuracy of the formula; for, if the heat absorbed by the vessel is allowed for, the discrepancy is increased. As we have already pointed out, however, the two formulæ are not on quite the same footing: Richmann's gives the ideal temperature that would be attained only if no heat whatever were lost to external sources on mixing, whereas Krafft's, by reason of the experimental determination of the constants, gives the temperature that would be observed in cases where no more precautions were taken to avoid loss of heat than Krafft himself took in his own experiments. That Richmann's experimental results are so much nearer the theoretical values than Krafft's

[1] See p. 13 above.

merely serves to emphasize the superiority of Richmann's *experiments*.

In Experiments III–VI Richmann did not make any allowance for the heat absorbed by the vessel. The sole purpose of these experiments was to demonstrate the superiority of the formula $\theta = \dfrac{m_1 t_1 + m_2 t_2}{m_1 + m_2}$ over that given by Krafft. Richmann was perfectly well aware that the temperature that would be observed on mixing m_1 measures of water at t_1° with m_2 measures at t_2° would not be exactly that given by the formula, but would be less by an amount corresponding to the heat absorbed by the vessel containing the cold water. Exact agreement between the calculated and observed temperatures would be against the correctness of a formula that gave the temperature of the mixture in terms of the masses and temperatures only: if, as in Krafft's formula, the agreement depended upon the presence of coefficients determined by experiment, then the formula could be applied with safety only when the mixing was carried out in the same vessel and under the same conditions as in the determination of the coefficients. Richmann realized that the correction for the vessel must be made by a separate calculation and could not be implicit in the formula. In testing his formula, he was therefore content to show that it gave temperatures that were consistently higher than those observed by amounts that could reasonably be ascribed to the heat absorbed by the vessel.

It would however have been a more convincing test of the accuracy of the formula, if Richmann had determined the "water-equivalent" of his vessel from one of the experiments (or better still, as the mean of a series of separate experiments) and had then calculated the correction for the vessel in each of the remaining experiments; by subtracting this correction from the temperature calculated from his formula he should have obtained the observed temperature. This is evidently what he did when he used Krafft's experimental results to test his own formula. In his previous memoir (*loc. cit.*, p. 158) he gave the following table:

Temperature calculated from Krafft's formula, $\theta = \dfrac{11m_1t_1 + 8m_2t_2}{11m_1 + 8m_2}$.	Temperature observed by Krafft.	Temperature calculated from Richmann's formula, $\theta = \dfrac{m_1t_1 + m_2t_2}{m_1 + m_2}$.	Temperature calculated from Richmann's formula, when vessel and thermometer taken as together equivalent to $1\frac{1}{2}$ measures of water.
68·12	68	76·00	68·15
86·28	85	92·00	87·20
98·73	98	103·14	99·88
108·73	107	112·25	110·00
115·75	113	118·66	115·62
120·42	118	122·90	121·48
124·37	123	126·54	125·46
128·52	128	130·42	129·59
132·80	130	134·46	133·80
134·34	134	135·85	135·61

Here Krafft (*Comment. Acad. Sci. Imp. Petrop.*, 1744–6, **14**, 230) had begun with 4 measures of water at 42° F., a measure being about $1\frac{1}{2}$ cubic inches. Into this he poured 1 measure of water at 212° F., and found the temperature of the mixture to be 68° F. He then added another measure of boiling water and found that the temperature of the mixture was 85° F. By repeated additions of 1 measure of boiling water he obtained the temperatures shown in the second column. The values in the last column were obtained by Richmann on the supposition that the vessel used by Krafft was equivalent to 1 measure of water and the thermometer to half a measure. Krafft made no reference whatever to his vessel, so that it was Richmann himself who arrived at the $1\frac{1}{2}$ measures as its equivalent in terms of water. This he evidently did by finding, from the data of the first experiment given in the table, how many measures of water at 42° F., mixed with 1 measure at 212° F., would give, according to his formula, a temperature of 68° F. If we put $\theta = 68$, $m_2 = 1$, $t_1 = 42$, and $t_2 = 212$ in $\theta = \dfrac{m_1t_1 + m_2t_2}{m_1 + m_2}$, we obtain the value 5·54 for m_1. If we now take $m_1 = 5\frac{1}{2}$, the formula gives the value 68·15 for θ. Richmann could therefore con-

sider that the 4 measures of water and the vessel at 42° F. gave
the same temperature on addition of 1 measure of boiling water
as would have been given by $5\frac{1}{2}$ measures of water at 42° F.
and no vessel—that is, the vessel (including the thermometer)
was equivalent to $1\frac{1}{2}$ measures of water.   Having thus arrived
at the "water-equivalent" of the vessel, Richmann could now
calculate, from his own formula, what temperatures Krafft
should have observed in each of the remaining experiments.

It is perhaps unfortunate that Richmann's failure to arrive at
a law of cooling from the first two experiments diverted his
researches into a fresh channel.   Had he now given his full
attention to the problem of the vessel, there can be little doubt
that his ability and insight would have led him, sooner or later, to
results that would have profoundly affected the history of the
discovery of specific heat.

H. Boerhaave (1732).   In this chapter we have so far
omitted in our discussion of the temperature of a mixture all
reference to Boerhaave.   He, in order of time, should come
before Krafft; for the *Elementa Chemiæ*, in which he discusses
the matter, was published in 1732.   Boerhaave, however, seems
to have been guilty of what we can only look upon as a serious
slip.   He says (*Elementa Chemiæ*, Leyden, 1732, 1, 268) that
if two equal quantities of the same liquid at different temperatures are mixed together, they will come to the same temperature, which is *half the excess of the hotter over the colder*; thus
1 pound of water at 32° F., mixed with 1 pound of water at
212° F., would give a temperature of 90° F., which is half
180 degrees, the difference between 32 and 212.   Boerhaave
can hardly have given much thought to a statement which, taken
as it stands, makes equal quantities of water, one at 50° F., the
other at 60° F., give on mixing a temperature of 5° F. !   It is
hardly possible to think otherwise than that he *meant* to say
that the temperature of the colder water would be *increased* by
this difference.   Yet throughout his discussion, whenever he
has occasion to refer to the temperature of a mixture, he makes
this same curious mistake.   The only possible explanation
seems to be that Boerhaave did no experiments on the subject,

but simply worked the whole thing out on paper, and so did not notice the absurdity to which he was committing himself. The roughest experiment must inevitably have drawn his attention to the error: for with equal quantities of water at 32° F. and 212° F., he would, like Krafft, have obtained a temperature somewhere in the neighbourhood of 108° F., certainly not 90° F. Shaw's translation of the *Elementa* is misleading on this point; for he makes Boerhaave say: "I made the following experiment," and goes on to describe the mixing of equal quantities of water at 32° F. and 212° F. (P. Shaw, *A New Method of Chemistry*, London, 1741, 1, 290.) Boerhaave, however, says nothing of the kind: he merely states what *would* happen if the experiment were made. It is suggestive that later on, in describing the well-known experiments on water and mercury, he says that Fahrenheit made these experiments for him. Boerhaave does not appear to have been an ardent experimenter as far as heat was concerned. He actually records that in this experiment "there appears something very subtile, *viz.* that the common degree of heat is lost, and the excess distributed equally between the two masses" (*New Method etc.*, I, p. 291). Boerhaave's account was described by Black as

so contrary to the fact, and so absurd in itself, that it is plain he had misunderstood Fahrenheit's account of what happened. He says, that whatever heat is in the colder water, that heat, and an equal quantity of the heat of the hot water, is extinguished, and disappears, and that the excess only of what is contained in the hot above the cold, remains in the mixture, and is equally distributed through it. If this were true, it would be easy, by adding hot water, to produce a mixture colder than the cold water. Boiling water, added to water a little less hot, would produce a mixture colder than ice. Dr. Boerhaave's account of what happens in this experiment must therefore be disregarded (p. 60).

Yet Crawford, quoting "nearly in the words of the author," stated that Boerhaave added equal quantities of water at 32° and 212° and found that the temperature of the mixture was 122° (*Animal Heat*, 1779 edn., p. 10; 1788 edn., p. 86).

Johan Carl Wilcke [1] **(1772).** There is a widespread idea that Wilcke's researches were the result of information supplied by Swedish students who had attended Black's lectures. But according to Wilcke's own statement, made at the beginning of his memoir on the cold produced by snow in melting, *Om Snöns kyla vid Smältningen* (*K. Svenska Vet. Akad. Handl.*, 1772, 33, 97),[2] his experiments were the result of pure chance. Early in the year 1772 [3] there had been a heavy fall of snow, and Wilcke, with the commendable intention of being helpful, set about the task of melting the snow that had accumulated in a small court-yard. He poured hot water on the snow, expecting that the hot water would convert many times its weight of snow at a temperature slightly below 0° C.[4] into water at a temperature slightly above 0° C. The complete lack of success with which his efforts met led him to the conclusion that "some unknown factor was here involved, and that in the melting of snow the same laws are not followed as in the distribution of heat between bodies, especially water."

---

[1] The correct Swedish spelling is Wilcke: the spelling Wilke, so often met with, is not used by the Swedish writers, but only by the Germans. It seems a pity that any but the correct Swedish form of the name should have crept into use in England and elsewhere.

[2] The account that follows is based entirely on the original Swedish memoirs. The German versions by Kästner referred to on page 11 (foot-note) have been compared with the originals and found, with a few un-important exceptions, to be faithful translations.

[3] Wilcke himself thus sets at rest any doubts there may have been as to the exact date when he first turned his mind to a subject that had already occupied Black's attention for so long—it was in the early part of the year 1772.

[4] Throughout his work, Wilcke, unlike the preceding workers we have discussed, used the Swedish or so-called Celsius thermometer, now known as the Centigrade thermometer. Anders Celsius, Professor of Astronomy at Upsala, proposed in 1742 a scale of temperature in which the M.P. of ice was taken as 100° and the B.P. of water at standard pressure as 0°. Strömer of Upsala inverted this scale in 1749, and the Swedish thermometer was thereafter graduated with 0° for the M.P. of ice and 100° for the B.P. of water at standard pressure. Meanwhile, Christin of Lyons had inverted Celsius's scale in 1743, but disregarded barometric pressure.

The particular law of distribution of heat between bodies to which Wilcke is here referring is Richmann's Law:

Temperature of the mixture $= \dfrac{m_1 t_1 + m_2 t_2}{m_1 + m_2}$, where m_1, m_2

are the masses, t_1, t_2 the corresponding temperatures. Wilcke had apparently fully satisfied himself as to the truth of this law for two quantities of water, even in the case where one was ice-cold. "Snow," he says, "which is nothing but frozen water, should obey the same law, cooling the warmer water and removing from it a proportionate amount of heat." Experiment, however, gave an entirely different result. While ice-cold water, mixed with an equal quantity of water at 68° C., gave the temperature of 34° C. required by Richmann's formula, an equal quantity by weight of snow not only reduced the temperature of the water to 0° C., but also left some of the snow unmelted at the end. Repeated experiments with different proportions of snow to water gave the same result: in every case a considerable quantity of heat was lost. "I therefore sought," he says, "to discover the laws of this behaviour, or some rule by which the temperature of the mixture could in all cases be determined" (*loc. cit.*, p. 99).

To this end [continues Wilcke] I provided myself with a sensitive balance, a good thermometer, and also several tall, thin, cylindrical vessels made of sheet iron (*järnbleck*) in which the snow and the water could be weighed out and the snow afterwards poured over at once into the water. The vessels were set on three vertical legs, or on a fine edge, to reduce the loss of heat. For the same reason I seldom used less than 1 pound, usually 3 to 5 pounds, of snow-water in an experiment. In experiments of this type it is impossible to avoid all sources of error: the temperature of the vessel and air, evaporation of the water, the taking up and putting down of the thermometer, the slightest spilling, the time taken, but above all the difficulty in always procuring melting snow in the same state—if sufficient attention is not paid to these points, marked errors can be introduced, though I seldom found them to amount to 2 degrees, and then the cause was not far to seek.

Wilcke's method of procedure was as follows. He took a

definite mass of melting snow and added it to 1, 2, 3, 4, etc. times its mass of water at a known temperature (shown in Column A in the Table). The temperature of the mixture was observed (Column B). The next step was to determine, by calculation from Richmann's formula, the temperature of such a mixture when ice-cold water was used instead of the snow (Column C). The lowering of the temperature of the mixture caused by the substitution of snow for ice-cold water Wilcke regarded as a measure of the loss of heat due to the snow (Column D). For example, when melting snow was added to an equal mass of hot water at 98° C., the temperature of the mixture was observed to be $13\frac{1}{2}$° C. When ice-cold water was used instead of melting snow, the temperature of the mixture is 49° C. (the value given by Richmann's formula). Hence the snow was responsible for a loss of $(49 - 13\frac{1}{2})$ or $35\frac{1}{2}$ degrees. Wilcke found that as long as the proportion of snow to water remained the same, no matter what the actual masses were, this difference was constant and independent of the temperature of the hot water. He therefore took the mean of all the values obtained for any given proportion of snow to water and used this mean in subsequent work.

TABLE I

1 part of snow to 1 part of water

A Temp. of Hot Water	B Temp. of Mixture (snow).	C Temp. of Mixture (ice-cold water).	D Loss of Heat.
$98 +$	$13\frac{1}{2} +$	$49 +$	$35\frac{1}{2} +$
89	$7\frac{1}{2}$	$44\frac{1}{2}$	37
$81\frac{1}{2}$	$5\frac{1}{2}$	$40\frac{3}{4}$	$35\frac{1}{4}$
84	6	42	36
78	3	39	36
77	2	$38\frac{1}{2}$	$36\frac{1}{2}$
75	1	$37\frac{1}{2}$	$36\frac{1}{2}$
70	0		
		Mean	$36\frac{3}{28}$

TABLE I (*continued*)

1 part of snow to 2 parts of water

A Temp. of Hot Water	B Temp. of Mixture (snow).	C Temp. of Mixture (ice-cold water).	D Loss of Heat.
92	37	$61\frac{1}{3}$	$24\frac{1}{3}$
86	$32\frac{3}{4}$	$57\frac{1}{3}$	$24\frac{5}{12}$
83	31	$55\frac{2}{3}$	$24\frac{1}{3}$
$75\frac{1}{2}$	$26\frac{1}{3}$	$50\frac{1}{3}$	24
72	24	48	24
70	$21\frac{1}{2}$	$46\frac{1}{3}$	$25\frac{1}{6}$
65	19	$43\frac{1}{3}$	24
63	$18\frac{1}{2}$	42	$23\frac{1}{2}$
51	11	34	23
45	$6\frac{1}{2}$	30	$23\frac{1}{2}$
34	$0\frac{1}{2}$	24	
20	$0\frac{1}{4}$		
		Mean	$24\frac{1}{40}$

With 3 parts of water to 1 of snow, the mean was 18
  „ 4 parts of water to 1 of snow, the mean was $14\frac{3}{10}$
  „ 5 parts of water to 1 of snow, the mean was $12\frac{1}{8}$
  „ 6 parts of water to 1 of snow, the mean was $10\frac{3}{8}$
  „ 7 parts of water to 1 of snow ⎱ For these proportions only one experi-
  „ 8 parts of water to 1 of snow ⎰ ment was made and no mean given.

Wilcke next took 1 pound of water and added to it successive equal masses of melting snow, each addition being 4 half-ounces. The first mixture thus contained 8 parts of water to 1 of snow, the second 9 parts of water to 1 of snow (as the 4 half-ounces of snow had melted and so increased the quantity of water by 4 half-ounces), and so on up to 16 parts of water to 1 of snow. The table at head of p. 82 shows the results obtained.

Wilcke then repeated this experiment, but began with 2 pounds of water at 78° C. instead of 1 pound. The mass of snow added each time was again 4 half-ounces. In this way he obtained proportions of water to snow from 16 parts of water to

Water : Snow.	A Temp. of Hot Water.	B Temp. of Mixture (snow).	C Temp. of Mixture (ice-cold water).	D Loss of Heat.
8 : 1	78	61	$69\frac{1}{3}$	$8\frac{1}{3}$
9 : 1	61	47	$54\frac{9}{10}$	$7\frac{9}{10}$
10 : 1	47	36	$42\frac{8}{11}$	$6\frac{8}{11}$
11 : 1	36	$26\frac{1}{2}$	33	$6\frac{1}{2}$
12 : 1	$26\frac{1}{2}$	$18\frac{1}{2}$	$24\frac{6}{13}$	$5\frac{25}{26}$
13 : 1	$18\frac{1}{2}$	12	$17\frac{1}{7}$	$5\frac{1}{7}$
14 : 1	12	$6\frac{3}{4}$	$11\frac{1}{5}$	$4\frac{9}{20}$
15 : 1	$6\frac{3}{4}$	2	$6\frac{21}{64}$	$4\frac{21}{64}$
16 : 1	2	$0\frac{1}{4}$	$1\frac{15}{17}$	$1\frac{43}{68}$

1 of snow up to 32 parts of water to 1 of snow. In this case he gives the temperatures of the mixture only, as follows:

Water : Snow.	Temperature of Mixture.
16 : 1	68
17 : 1	59
18 : 1	44
19 : 1	38
20 : 1	32
21 : 1	28
22 : 1	24
23 : 1	20
24 : 1	16
25 : 1	$13\frac{1}{8}$
26 : 1	11
27 : 1	9
28 : 1	6
29 : 1	3
30 : 1	1
31 : 1	0
32 : 1	0

This last table gives all the data necessary, as the temperatures given in column C in the first tables are calculated from Richmann's formula, and the loss of heat given in column D then found by subtraction. Wilcke does not, however, make any use of the data for proportions higher than 8 : 1 in his subsequent argument.

For convenience in following the next and most important

step in the argument, we will set out here the mean values for the loss of heat found in the first eight experiments.   Thus:

Water : Snow.	Loss of Heat in degrees.
1 : 1	$36\frac{3}{28}$
2 : 1	$24\frac{1}{40}$
3 : 1	18
4 : 1	$14\frac{3}{10}$
5 : 1	$12\frac{1}{8}$
6 : 1	$10\frac{3}{8}$
7 : 1	—
8 : 1	—

Referring to these values, Wilcke says:

I at once perceived that the numbers found decreased in a Harmonic Progression whose terms are inversely as the terms of the Arithmetic Series 1, 2, 3, 4, . . . and are thus respectively $\frac{1}{2}$, $\frac{1}{3}$, $\frac{1}{4}$, $\frac{1}{5}$ . . . of some definite and fixed unit, which can be found through the means of the numbers, since each one of the observed losses should be $\frac{1}{2}$, $\frac{1}{3}$, $\frac{1}{4}$, $\frac{1}{5}$, etc. of this unit (*ibid.*, p. 103).

He then gives the following table:

A	B	C	D	E		
Water : Snow.	Heat lost (degrees).	Heat lost (unit).	"Unit" itself calculated.	Heat lost in degrees calculated from unit.		
				72	73	74
1 : 1	$36\frac{3}{28}$	$\frac{1}{2}$	$72\frac{3}{14}$	36	$36\frac{1}{2}$	37
2 : 1	$24\frac{1}{40}$	$\frac{1}{3}$	$72\frac{3}{40}$	24	$24\frac{1}{3}$	$24\frac{2}{3}$
3 : 1	18	$\frac{1}{4}$	72	18	$18\frac{1}{4}$	$18\frac{1}{2}$
4 : 1	$14\frac{3}{10}$	$\frac{1}{5}$	$71\frac{1}{2}$	$14\frac{2}{5}$	$14\frac{3}{5}$	$14\frac{4}{5}$
5 : 1	$12\frac{1}{8}$	$\frac{1}{6}$	$72\frac{3}{4}$	12	$12\frac{1}{6}$	$12\frac{1}{3}$
6 : 1	$10\frac{3}{8}$	$\frac{1}{7}$	$72\frac{5}{8}$	$10\frac{2}{7}$	$10\frac{3}{7}$	$10\frac{4}{7}$
7 : 1	—			9	$9\frac{1}{8}$	$9\frac{1}{4}$
8 : 1	—	Mean value $72\frac{1}{6}$		8	$8\frac{1}{9}$	$8\frac{2}{9}$

This table seems to call for some explanation.   Column B gives the mean value for the loss of heat determined by the experiments for the proportions of water to snow shown in column A.   These heat losses are all very nearly aliquot parts of the same number, 72, which Wilcke calls the *Enhet*, i.e.

"unit." The numbers in column D are obtained by multiplying the corresponding number in column B by the denominator of the corresponding fraction in C, and give the value of the "unit" as determined by each set of experiments. The mean of this column, $72\frac{1}{6}$, is the best value for the "unit" given by the whole set of experiments. Wilcke did, however, carry out another series of experiments, in which he obtained the value 74. In column E he therefore shows the loss of heat as calculated from column C and each of the possible values of the "unit," 72, 73, or 74.

Wilcke keeps to the value 72 in preference to the mean value 73, as it is an easier number to use in the subsequent argument, and, as he says, the precise number is unimportant as far as the theoretical discussion is concerned. The important point is that such a number exists; later experiments will determine its exact value.

In arriving at this number, 72, which is, after all, the culmination of his experiments, Wilcke chooses to be anything but direct and explicit. A first reading produces a sense of disappointment and creates a suspicion that there has been a certain amount of juggling with mere figures. Why bring in a totally unnecessary harmonic progression? When the fact is pointed out, it is seen that the numbers do indeed form such a series, though probably few besides Wilcke would "at once perceive" it. To arrive at so important a number merely by finding the first term of a harmonic progression seems highly unsatisfactory; the number obtained in this manner seems somehow fortuitous, and the reader is left groping for some underlying physical reality.

Yet there is no occasion for obscurity. If we take the heat losses to be 36, 24, 18, $14\frac{1}{2}$, 12, $10\frac{1}{2}$, it is quite readily seen that, with two exceptions, these numbers are all exact submultiples of the number 72; that is, they are, in order, $\frac{1}{2}, \frac{1}{3}, \frac{1}{4}, \frac{1}{5}, \frac{1}{6}, \frac{1}{7}$ of 72. And these same fractions give the proportion of melted snow in the mixtures: thus, 1 part of snow to 5 parts of water gives a mixture in which one-sixth is melted snow, and one-sixth of 72 is the loss of heat due to this snow. If the mixture consisted

PLATE V.

JOHAN CARL WILCKE (1732–96)

Studied at Göttingen and Rostock. Professor of Experimental Physics at the Military Academy, Stockholm, and Secretary (1786–96) of the Swedish Academy, to which he contributed numerous memoirs on various branches of physics.

(Reproduced by kind permission of the Royal Swedish Academy of Sciences.)

wholly of melted snow at 0° C., the loss of heat would be the full 72 degrees.   This number 72 therefore represents the heat lost in converting snow at 0° C. into water at 0° C.   It is not strictly, however, the latent heat.   It is a loss of temperature rather than a loss of heat; and it is entirely independent of the unit of mass chosen.   Throughout Wilcke's argument, it does not matter what the actual masses are, provided they bear a definite ratio to each other: whether 1 pound of snow is added to 2 pounds of water or 2 pounds of snow to 4 pounds of water, the loss of heat will still be one-third of 72 degrees Centigrade.

The value 72 is not, of course, a good one.   The principle on which the calculation of the "unit" is based is perfectly sound and should give a value nearer 80, if the data are themselves correct.   A consideration of the first set of experiments in table I (pp. 80–1 above) shows that, in order that the "unit" should be 80, the mean loss of heat should be 40: that is, the values shown under D for the loss of heat in each experiment are too low.   These values are obtained by subtracting the temperatures of the hot water and snow mixture (col. B) from the corresponding temperatures of the hot water and ice-cold water mixture (col. C).   In order that the values in column D should be higher than those shown, either the values in column B should be lower or those in column C higher.   A very likely source of error that would lead to too high values under B lies in the use of snow in the experiments; if any of the snow had melted before it was added to the hot water, the cooling effect of the resulting ice-cold water would be considerably less than that of the same quantity of snow.   Wilcke, however, seems to have been fully aware of this source of error; for he specially emphasizes the importance and difficulty of obtaining the snow in just the right condition (see p. 79 above)—evidently it did not occur to him to use crushed ice on the point of melting in place of snow.   The good agreement among the values in column D seems, however, rather against this as the explanation of the low value obtained, especially as there is a very good reason for concluding that the calculated temperatures shown in column C are really all too low. At the very beginning of the memoir Wilcke states that when ice-cold water was mixed with an equal quantity of water at 68° C., the temperature of the mixture was *observed* to be 34° C., which, it will be seen, is exactly the value required by Richmann's formula. This hardly seems possible: Richmann's formula gives the ideal, unattainable value for the temperature when no heat whatever is lost. Wilcke certainly took steps to avoid loss of heat due to conduction; but he appears to have made no allowance at all for the heat capacity of the vessel, and this for sheet-iron vessels holding at least 2 pounds of water must have been considerable. The values shown in column C are all exactly those required by Richmann's formula, that is, half the temperature of the hot water in the first set of experiments, two-thirds in the second set, and so on. They were evidently calculated, not determined by independent experiments, though Wilcke states (*ibid.*, p. 99) that they were determined from Richmann's formula and experiment. If Wilcke had actually added the ice-cold water to the hot water, he would certainly have observed higher temperatures than those given; for in addition to the hot water there was also an iron vessel at the same temperature contributing in no small degree to the temperature of the mixture.

It is interesting to see what effect any allowance for the vessel would have had on the value of the unit. This we can do by assuming some reasonable value for the mass of the vessel and including it in the subsequent calculations. In the experiments (p. 80 above) we do not know the actual masses of water and snow used, but only that the mass of the water was equal (in the first set of experiments) to the mass of the snow added. We cannot therefore assign a definite value to the mass of the vessel, but must express it in terms of the mass of the water. The simplest is to assume that the mass of the vessel was equal to that of the water. Since the specific heat of sheet-iron, the material of which the vessel was made, is about $\frac{1}{9}$, we can say that, in the first set of experiments, water : snow = $1\frac{1}{9}$: 1. With these proportions, snow at 0° C., when added to the water at 98° C., gave an observed temperature of $13\frac{1}{2}$° C. The next thing is to calculate from Richmann's formula the temperature that would have been obtained, had ice-cold water been used instead of snow.   Putting $m_1 = 1$, $t_1 = 0$, $m_2 = 1\frac{1}{9}$ and $t_2 = 98$ in the expression $\dfrac{m_1 t_1 + m_2 t_2}{m_1 + m_2}$, we obtain the value $51\frac{1}{2}°$ (nearly) for the temperature of the mixture.   The loss of "heat" due to the substitution of snow for the ice-cold water was therefore $(51\frac{1}{2} - 13\frac{1}{2})$ or 38 degrees.   In arriving at the "unit" from this loss of heat, it is necessary to remember that the proportion of water to snow is no longer 1 : 1, but $1\frac{1}{9}$: 1, so that the "unit" is no longer obtained by doubling the loss of heat, but by multiplying this loss by 19/9, that is, the "unit" is $38 \times 19/9$ or $80\frac{2}{9}$.

We must not attach too much significance to the fact that in this way we have arrived at a value for the "unit" that closely approaches the value now accepted for the latent heat of fusion of snow.   All that we are entitled to say is that, if the experimental data are to be relied on, then the vessel used in this experiment did weigh very nearly as much as the water.   As, however, we do not know the actual mass of the water, this information is not of much use to us.   If we make the same calculations for some of the other experiments, we obtain values for the "unit" that are usually in the neighbourhood of 80, though sometimes too high, sometimes too low.   The second experiment, for instance, gives the value 83.   These inconsistencies do not reflect in any way on Wilcke's experiments: they arise simply because we take the mass of the vessel as equal to that of the water in each case, and the results of our calculations show that we were wrong in doing so—that Wilcke used a different vessel or a different quantity of water in the same vessel, in either case making the ratio of the masses no longer one of equality.

To return to Wilcke's work, the conclusion that he draws from the experiments is that

the heat in a mixture of water and snow is not divided without loss, remaining active, as is the case in a mixture of water and water, but a definite and fixed quantity of it, corresponding to 72 degrees on the thermometer, is always lost. This heat the snow, which in this respect can be held to be in the negative state, draws to itself or destroys, without becoming warmer as a result, but merely becoming liquid. Thus this 72 degrees serves only to produce liquefaction. Afterwards the melted snow behaves like ice-cold water, and any heat over the 72 degrees is uniformly distributed through the whole mass (*ibid.*, p. 105).

Thus if water at 72° is mixed with an equal mass of snow, all the heat is lost and the mixture is ice-cold. If the water is at 98°, 72 degrees of heat are lost and the remaining 26 degrees are divided equally between the two masses, making the temperature of the mixture 13°. If 3 parts of water at 40° are mixed with 1 part of snow, 72 degrees of heat are lost, and the remaining (120 − 72) or 48 degrees are divided equally among the four masses, making the temperature of the mixture 12°.

If the mass of the snow is 1, the mass of the water m, and the temperature of the water t, then the temperature of the mixture is given by

$$\frac{mt - 72}{m + 1}.$$

If the mass of the snow is put equal to n in all cases to avoid fractions, then the temperature of the mixture is

$$\frac{mt - 72n}{m + n}.^{[1]}$$

Wilcke then carried out several experiments to verify the formula, the results being shown in the following table.

[1] The purpose served by the introduction of n is not at once clear. The change is, however, a very trifling one and merely serves, as Wilcke says, to avoid fractions. Actual masses never enter into Wilcke's calculations, but always the ratio of the masses. If the mass of the water is an exact multiple of the mass of the snow, then the first expression is the one to use. But if the mass of the water is $1\frac{1}{2}$ times that of the snow, it is obviously easier for the purposes of calculation to treat the ratio as 3 : 2 rather than $1\frac{1}{2}$: 1.

A	B	C	D
Water : Snow.	Temperature of Water.	Observed Temp. of Mixture.	Calculated Temp. of Mixture.
3 : 2	67	11	$11\frac{2}{5}$
3 : 2	86	23	$22\frac{4}{5}$
4 : 3	$89\frac{1}{4}$	$18\frac{3}{4}$	$20\frac{1}{7}$
7 : 4	72	$20\frac{1}{4}$	$19\frac{9}{11}$
$41\frac{1}{2}$: 32	88	$18\frac{1}{2}$	$18\frac{1}{3}$

The close agreement between the observed and calculated values satisfied Wilcke that the formula was correct. We can see that it is so by putting θ for the temperature of the mixture and then writing the equation as

$$72n + \theta n = m(t - \theta),$$

in which the left-hand expression is clearly the heat necessary to melt n grams of snow at $0°$ C. and raise the temperature of the melted snow to $\theta°$ C. (assuming, of course, that 72 is the latent heat of fusion of snow), and the right-hand expression is the heat given up by m grams of water in cooling from $t°$ C. to $\theta°$ C.

Wilcke then goes on to make some general observations on the formula.

The temperature of the mixture will be $0°$ whenever

$$mt = 72n, \text{ (or } m : n = 72 : t).$$

Hence if to unit mass of snow (i.e. if $n = 1$), 1, 2, 3, etc. times its mass of water is added, then in order that the mixture may be ice-cold water, that is, all the snow melted and all the heat lost, the temperature of the water added must be 72, 36, 24, 18, etc. degrees respectively. It can therefore be concluded that, since any heat over and above these is not taken up by the snow but is left in the mixture, water at $72°$ contains twice as much heat as the same mass of water at $36°$, three times as much as the same mass of water at $24°$, and so on. Thus the quantity of heat in a given mass of water is directly proportional to the temperature of the water. In the thermometer, degrees correspond to equal increments in the volume of the mercury, so that it is clear that these increments are directly proportional to the quantity of that which we call heat in the water (*ibid.*, p. 107). If the state of melting snow or ice is regarded as a state of no heat at all (absolute deprivation, *någon absolut brist*), it follows that the degrees on the thermometer measure the actual heat present or are proportional to it.

If $mt > 72n$, then the mixture possesses a degree of heat $= \dfrac{mt - 72n}{m + n}$, and if $mt = 72n$, then the mixture is ice-cold. But if $mt < 72n$, then the expression remains negative. This last case puzzles Wilcke: it seems to indicate that the temperature is below the freezing-point, so that the water should freeze. This never happens; for however much melting snow is added, the mixture remains at the freezing-point. "It appears therefore that this heatless state in snow is something absolute (*något absolutum*), its sole effect being to produce a melting of the snow; it is not, like cold, something opposed to heat, or the negative of heat, but is really o" (*ibid.*, p. 108). The term "cold of snow," as the negative of heat, seems to Wilcke to be applicable to snow cooled below o° C. Ice-cold water is soon frozen by the addition of such snow.

This illustrates very well the kind of difficulty that might arise out of the confusion between temperature and quantity of heat, out of the reduction of all heat exchange to a matter of temperature alone. To Wilcke and those of his contemporaries and predecessors who held heat to be a substance, the notion of "quantity of heat in a body" was, of course, definite enough— simply "how much of this substance does the body contain?"; while "temperature of a body" was no less definite—"to what division on the thermometer does the mercury rise when placed in contact with the body?" The problem was to correlate the two, so that the thermometer could be used to measure quantity, and here they met with endless difficulties and seeming contradictions, as was inevitable while attention was fixed solely on the quantity of heat *in* a body instead of on the quantity of heat taking part in an exchange. As a result, temperature is still the only determinate quantity to Wilcke, so that when his expression comes out negative, he is faced with the difficulty of finding an explanation for a negative *temperature* that he knows to be impossible. If he had based the discussion in the last paragraph on *quantity* instead of on *temperature*, he would not have got into difficulties. And he was in a position to do this, for he had just deduced that 3 pounds of water at 24° will melt as much snow as 2 pounds at 36° or 1 pound at 72°, and so on, and as he had found that the heat necessary to melt 1 pound of snow (or any quantity of snow) always "corresponds to 72 degrees of heat on the thermometer," it is surprising that he did not come to the conclusion that the quantity of heat given up by the water to the snow is equal to the mass of the water multiplied by its temperature, taking, if he liked, the heat needed to melt 1 pound of snow at 0° C. as unit quantity of heat. He could then have reasoned: in all cases $mt - 72n$ is the quantity of heat available for raising the temperature of the mixture above 0° after $72n$ units have been used up in melting the snow. This, divided by $m + n$, the total mass of the water and melted snow, gives the resulting temperature.

Then, if $mt > 72n$, there is heat available for raising the temperature of the mixture by an amount $\dfrac{mt - 72n}{m + n}$ above 0° C.;

and, if $mt = 72n$, no such heat is available, and the mixture remains at 0° C.;

but, if $mt < 72n$, there is not enough heat available to melt all the snow. As $72n$ units are required to melt all the snow and only mt units were available, $72n - mt$ gives the amount of heat still needed to melt the rest of the snow, the mixture of snow, melted snow, and water being at 0° C.

Wilcke next considers the question as to whether the presence of water is essential: whether snow in melting takes up 72 degrees of heat without becoming warmer when the heat is derived from other bodies besides hot water.

To this end he placed a known mass of melting snow in one

of his sheet-iron vessels, and an equal mass of ice-cold water in a second vessel identical with the first. These two vessels were placed simultaneously in a saucepan of boiling water over the fire. As soon as the thermometer in the water reached 72° C., the vessel containing the snow was removed from the heater and suspended in the air. The thermometer in it at once sank to 2° C., and then, as some of the snow remained undissolved,[1] to 0° C., where it remained stationary. Here no water, except its own, had come in contact with the snow. The two equal masses, snow and water, had, in the same time, received the same amount of heat and should have reached 72° C. together. But when the water had reached this temperature, the snow did not appear to have received the slightest heat, but had merely become liquid. The sole effect of the 72 degrees of heat was to melt the snow, without any effect on the thermometer, just as when water at 72° C. is mixed with snow to give a mixture of ice-cold water.

Wilcke concludes this part of his memoir with the observation that, as snow in melting takes up 72 degrees of heat without becoming warmer, so water in freezing must give up 72 degrees of heat; and if this heat is imparted to another equal mass of ice-cold water, it will produce 72 degrees of heat in the latter, without its being possible to detect the loss of heat in the ice that was produced by the freezing of the original mass of water. This, Wilcke remarks, appears paradoxical, but not more so than that ice in melting should take up 72 degrees of heat without becoming warmer. His experiments on this matter were not yet complete. It would be interesting to know what form experiments took which aimed at heating one quantity of ice-cold water by freezing another! As, however, there seems to be no further reference to them in later memoirs, it may be assumed that Wilcke had but little success with them. It is greatly to his credit that he did look at the problem before him from every angle, and that he did not reject a conclusion simply on the grounds that it appeared paradoxical. Had he but

---

[1] Owing to the fact that the heating was discontinued at 72° C. instead of at 80° C.

realized the significance of the failure of these experiments, that the heat of the colder body is not available for warming the hotter body, he might even have been led to a discovery far greater than that which he had already made—the Second Law of Thermodynamics.

Wilcke devoted the rest of the memoir to considerations of a general character on the nature of heat. He says (*ibid.*, p. 111) that, while opinion is divided as to the ultimate cause of heat and cold, the most widely accepted view is that heat is a subtle fluid substance, which can be known only through the effects it produces. Wilcke calls it "fire" (*Eld*) or "fire-stuff" (*Eld-ämne*). He also makes the assumption that both solids and liquids exert an attraction on this substance, or rather their solid particles do so. It penetrates the pores of bodies and surrounds and adheres to their outer particles. At the freezing-point the particles of water are surrounded by a definite quantity of the substance, which keeps them apart and maintains their fluidity; if more is added, the volume of the water increases, while the presence of the additional "fire" is shown by a rise of the thermometer. On the other hand, a smaller quantity causes the water particles to come into immediate contact, so that their surfaces combine to form a solid—ice. Wilcke illustrates his ideas of these water particles by comparing them with a number of small, thin discs placed in water. While there is plenty of water present, discs lying on top of one another can easily be separated; but if the water is poured off, the discs adhere together with considerable force, becoming separated again when more water is added. The 72 degrees of heat absorbed by snow merely in becoming liquid correspond to the quantity of water that, by getting between the discs, is just sufficient to separate them.

In concluding this account of Wilcke's experiments on the heat taken in by snow in melting, it is almost unnecessary to add any comment about the superiority of Black's work, theoretical and practical, or to point out that Black, unlike Wilcke, extended his ideas and experiments to include the latent heat of vaporization. But it must be emphasized that Wilcke's experiments were carried out in the early part of 1772, that is, more than ten years after Black's determination of the latent heat of fusion of ice (see p. 31 above): and that Robison's claim (see p. 36 above) that the discovery of latent heat was "the undivided property" of his "ingenious and acute preceptor" is unassailable.

CHAPTER IV

WILCKE'S WORK ON SPECIFIC HEAT

Wilcke begins his memoir on the specific heat of solids and its measurement, *Rön om Eldens specifica myckenhet uti fasta kroppar och des afmätande* (*K. Svenska Vet. Akad. Nya Handl.*, 1781, 2, 49), with a comment on the indifferent success that had so far attended all attempts to solve the problem of the quantity and distribution of heat in bodies. In this connexion he quotes a passage from Klingenstjerna's Comments on Musschenbroek's *Physica*,[1] in which Klingenstjerna disagrees with the conclusion reached by both Boerhaave and Musschenbroek that, when bodies are at the same temperature, heat is necessarily distributed uniformly among them, so that a cubic foot of gold contains as much heat as a cubic foot of air or feathers at the same temperature. According to Klingenstjerna, no conclusion can be drawn as to the distribution of the heat from the mere fact that the bodies are at the same temperature. The supposition that heat distributes itself among different substances in proportion to their densities is equally consistent with uniformity of temperature. For if, when the steady state is reached, the quantity of heat in water, for example, bears to the quantity of heat in quicksilver the same ratio as the density of the water bears to the density of the quicksilver, and so on in all substances, then no substance could cool or warm another, and so no substance would cool or warm the thermometer, and all would therefore still indicate the same temperature. Other phenomena, how-

---

[1] We have unfortunately been unable to trace this work by Klingenstjerna. It is probably a translation into Swedish of Musschenbroek's *Physica*, with Comments by Samuel Klingenstjerna (1689–1765), Professor of Mathematics at Upsala.

ever, suggest that, of two bodies at the same temperature, the denser contains more heat than the less dense. For example, two spheres of equal diameter and at the same temperature, when immersed in equal quantities of water at the same temperature, do not raise the temperature of the water by the same amount if the spheres are of different densities; thus, if one is of gold and the other of tin, the gold gives more heat to the water than the tin. Surely, then, the gold contains more heat than the tin? Again, of two cold bodies at the same temperature, the denser is colder to the touch than the less dense, so that a slab of marble possesses more cold than a piece of wood.

Wilcke himself considers that the difficulty in dealing with this problem of the relative quantities of heat in bodies is undoubtedly due to the lack of a convenient and reliable standard of measurement by which the relative, if not the absolute, amounts of heat in bodies can be determined (*loc. cit.*, p. 51). In his experiments on the cold of snow he had discovered the remarkable fact that melting snow absorbs a definite quantity of heat merely in becoming liquid.

This proved that "fire" [*eld*] or "heat" [*varme*] is really a substance whose quantity can be measured; its deficiency or excess changes a body's state from solid to liquid; it can be present in abundance in the body without being detected by the thermometer; it can, however, be set free again and become evident as heat [*varme*]; it thus produces all the phenomena of artificial heat [*hetta*] and cold (*ibid.*, p. 52).

Here, then, is a means of determining at least the relative quantities of heat in bodies. All that is necessary is to find "how much melting snow is used in cooling the different bodies from a definite temperature to the freezing-point." Since all the heat which the body loses is in the resulting snow-water, the quantity of heat may be known from the quantity of the water.

Experimental difficulties at once arose. Either the snow must be placed on the body or the body on the snow; and in either case the resulting snow water was absorbed so quickly by the rest of the snow that it was all but impossible to determine how much had been melted. Attempts to overcome this difficulty by taking successive small quantities of snow were not very successful, as the results were unreliable owing to the length of time required and the consequent loss of heat entailed. Wilcke then tried putting a known quantity of snow in a known quantity of ice-cold water, placing the body, previously heated to a known temperature,[1] in this mixture, and determining how much snow could be dissolved in this way without raising the temperature of the mixture above 0° C. or leaving any snow undissolved at the end. "This did succeed, but the difficulty of the operation and the unexpected results obtained soon led me to think of another way, easier, though less direct, of finding what I sought" (*ibid.*, p. 53). This method consisted in taking a quantity of ice-cold water whose mass equalled that of the body; in this the body was immersed, after having been previously heated to a known temperature, in particular to the temperature of 72° C.; and the temperature that resulted was observed. Wilcke then calculated from Richmann's formula the mass of water at the temperature of the body necessary to produce a mixture of the same temperature when added to an equal mass of ice-cold water. He then calculated from his own formula for the melting of snow what mass of snow would be required to reduce the temperature of the mixture to 0° C. This mass of snow could then be weighed out and put to the test either by adding it to the mixture or by placing it directly on the body.

This succeeded in every way, but I soon saw that the last operation with the snow was not really necessary, as the specific heat [*Specifica-varme*—this is the first time Wilcke uses the term], which was what I sought, could be found directly from the temperature of the water, from which the quantity of snow is calculated (*ibid.*, p. 54).

---

[1] Wilcke especially favours the temperature of 72° C., since, according to his earlier experiments, water at this temperature will just melt an equal mass of snow at 0° C. without raising its temperature. This particular temperature of 72° C. he refers to as the "standard temperature" (*likare-grad—likare*, gauge).

At this point Wilcke interrupts the account of his experiment in order to discuss "Specific Heat" and the law of distribution of heat in different bodies. It seems better, however, to continue here with the details of the experiments and to return to the conclusions drawn from them later.

The method of carrying out the experiments was as follows:

I. The body or metal whose specific heat was required was weighed on an accurate balance. Usually about 1 pound was taken.

II. The body was suspended by a thread in a tall metal [1] vessel filled with boiling water. A sensitive thermometer was placed in the vessel in contact with the body.

III. A quantity of ice-cold water was weighed in a very thin metal vessel attached by threads to the arm of the balance, the mass of this water being made equal to the mass of the body. The ice-cold water used was obtained from another vessel filled with melting snow, but in pouring off the water care had to be taken to prevent any of the snow from passing over, as this would have rendered the experiment worthless.

IV. The body was then quickly withdrawn from the boiling water and lowered into the ice-cold water, where it was left suspended, care being taken that it did not touch the sides or bottom of the vessel.

V. A sensitive thermometer, divided into $\frac{1}{4}$-degrees, was placed in the mixture and the temperature noted when the temperature of the water above and below the body was the same as that of the body.

VI. The experiment was repeated as often as the temperature of the water in the heater allowed. The specific gravity of the body was also measured in the same water.

---

[1] There is some doubt as to the metal of which the vessels were made in the present experiments. Those of nine years before are clearly described as being of sheet-iron (*järnbleck*). In the present case, the vessels are described as *bleck-dosor*, which in present-day Swedish means "tin boxes," though in 1781 it probably meant simply "sheet-metal boxes." In any case, the metal would be tinned sheet-iron, not pure tin. Both the German and French versions favour "tin."

The results of the experiments are set out in the following tables, in which:

Column A gives the temperature (t) of the metal on removal from the heater. The temperatures are all on the Swedish scale (i.e. Centigrade).

Column B gives the temperature (θ) of the mixture after the body was inserted and the temperature had become uniform.

Column C gives the temperature which the mixture would have had if water of the same mass and temperature had been used instead of the body. This is found from Richmann's **Law**, where the temperature (θ) is given by

$$\theta = \frac{m_1 t_1 + m_2 t_2}{m_1 + m_2},$$

where m_1, m_2 are the masses, t_1, t_2 the corresponding temperatures.

If the masses are equal and the water ice-cold, then $\theta = \dfrac{t_2}{2}$, that is, half the temperature of the hot water, and therefore half the temperature of the body (given in col. A).

Column D gives the temperature which the mixture would have had if water of the same *volume* and temperature had been used instead of the body.

The ice-cold water and the body both have the same mass; their volumes are therefore inversely proportional to their specific gravities. If g is the specific gravity of the body, m_1 its mass, and m_2 the mass of the water equal in volume to the body, then

$$m_2 : m_1 = 1 : g, \text{ or } m_2 = \frac{m_1}{g};$$

and if the mass of the cold water (and therefore of the body) is taken as unity, then $m_2 = \dfrac{1}{g}$.

Putting $m_1 = 1$, $t_1 = 0$, $m_2 = \dfrac{1}{g}$ and $t_2 = t$ in Richmann's formula gives $\theta = \dfrac{t}{g + 1}$ for the temperature of the mixture. This is how column D was calculated.

Column E shows how much of the hot water, which is at the same temperature (t) as the body, is needed in proportion to the mass of the body (which is equal to the mass of the ice-cold water) in order that, on being mixed with the ice-cold water, it may give the same temperature (θ) as the body gave.

Putting $m_1 = 1, t_1 = 0, t_2 = t$ in Richmann's formula gives

$$m_2 = \frac{\theta}{t - \theta},$$

from which column E is calculated.

This value, $\dfrac{\theta}{t - \theta}$, found for the mass of water, of the same temperature as the body, which gives to the mixture the same temperature as the body gave, expresses the specific heat of the substance in comparison with water; or the ratio of the quantity of heat in each particle of the body to the quantity of heat in each particle of water. For, if we neglect the slight difference due to inequalities in temperature, the mass and volume of the ice-cold water bear to the mass and volume of the hot water at t (col. A) that gives the same temperature to the mixture as the body the ratio $1 : \dfrac{\theta}{t - \theta}$. The number of material parts, or particles, in this water is in the same proportion. But the body is as heavy, and therefore has as many particles, as the ice-cold water. Therefore the number of particles in the body bears to the number of particles in the warm water the ratio $1 : \dfrac{\theta}{t - \theta}$. Both, however—the body and the warm water—give up just the same quantity of heat to the mixture with ice-cold water; and this, divided equally among its particles, makes for each particle a quantity of heat, or specific heat, which is inversely as the number of particles in the body, and in the water; that is, as $\dfrac{\theta}{t - \theta} : 1$ (*ibid.*, p. 60).

In the table, the values of $\dfrac{t - \theta}{\theta}$ are given (in decimals) instead of the reciprocals. This is the specific heat of the water particles when the specific heat of the substance of the body is taken as unity.

Column F shows the quantity of snow required to render the quantity of water at temperature t, given in column E, ice-cold, and therefore the quantity required to render the body itself, which is at the same temperature, ice-cold, that is, to remove all it heat.

By the formula found in the memoir of 1772, namely

$$\frac{mt - 72n}{m + n},$$

the mixture of snow and water will be ice-cold and all the snow dissolved when $mt = 72n$, where m is the mass of water at temperature t equivalent to the body, and n is the mass of the snow, that is, $72 : t = m : n$, or the mass of the snow (found) $= n = \dfrac{mt}{72}$. For m put $\dfrac{\theta}{t - \theta}$ (from the earlier columns) where t is the temperature of the body. Then the mass of the snow which, in melting, makes the body ice-cold

$$= n = \frac{t\theta}{72t - \theta} = \frac{t}{72} \times \frac{\theta}{t - \theta},$$

that is, the specific heat of the substance (from column E) multiplied by the temperature of the body (from column A) divided by 72.

If $t = 72$, then $n = \dfrac{\theta}{t - \theta}$, so that, if the initial temperature of the body is $72°$, the mass of the snow at once gives the specific heat of the substance. In the tables, however, only the denominators of the fractions are given (as decimals), showing how much more warm water is required to melt the snow when its mass is taken as unity.

Wilcke gives 14 tables of experimental results in all. Of these we have selected four, namely those for gold, lead, silver and glass, as sufficiently illustrating his work.

For convenience in reference, the quantities shown in the columns A, B, ... F are set out again here, with the symbols and expressions attached to them:

A. Temperature of heater, i.e. initial temperature of body . t

B. Temperature of mixture . . . . θ

C. Calculated temperature when water equal in mass and temperature is substituted for the body . . . . $\dfrac{t}{2}$

D. Calculated temperature when water equal in volume and temperature is substituted for the body . . . . $\dfrac{t}{g+1}$

E. Reciprocal of the mass of water equivalent to the body, i.e. $\dfrac{1}{\text{specific heat}}$ . . . . $\dfrac{t-\theta}{\theta}$

F. Reciprocal of calculated mass of snow required to destroy heat of body . . . . $\dfrac{72(t-\theta)}{t\theta}$

TABLE I

Gold. Sp. Gr. 19·040

A	B	C	D	E	F
73 +	3½	36½	3·642	19·857	19·585
62½	3	31¼	3·100	19·833	22·840
53	2½	26½	2·6	20·500	27·840
48	2¼	24	2·3	20·333	30·499
39½	2	18¾	1·9	18·750	32·910
35	1¾	17½	1·7	19·000	39·080
			Mean	19·712	

TABLE II

Lead. Sp. Gr. 11·456

A	B	C	D	E	F
86 +	3½ +	43	6·904	23·571	19·733
83	3¼	41½	6·583	24·538	21·285
74	3	37	5·940	23·666	23·026
73	3	36½	5·860	23·333	23·012
58	2½	29	4·656	22·200	—
55	2¼	27½	4·415	24·779	—
52½	2¼	26¼	4·214	22·333	—
42	1¾	21	3·371	23·000	—
34½	1½	17¼	2·769	22·000	—
			Mean	23·515	

TABLE III

Silver. Sp. Gr. 10·001

A	B	C	D	E	F
89	6½	44½	8·090	12·692	10·267
78	6	39	7·091	12·000	11·076
70	5½	35	6·363	11·727	12·062
63	4¾	31½	5·727	12·263	14·014
56	4¼	28	5·090	12·176	12·654
50	4	25	4·545	11·500	16·560
44½	3½	22¼	4·045	11·714	18·952
			Mean	12·010	

TABLE XIV

Glass. Sp. Gr. 2·386

A	B	C	D	E	F
86 +	12¾	43	25·398	5·725	—
73½	10¾	36¾	21·707	5·837	5·958
59½	8¾	29¾	17·572	5·771	4·769
50½	7½	25¼	14·914	5·733	—
42	6½	21	12·404	5·461	—
34½	6	17¼	10·188	4·750	—
30	5	15	8·860	5·000	—
24	4½	12	7·088	4·333	—
			Mean	5·326	

So far we have kept very closely to Wilcke's own account of his work, and made no attempt to curtail his argument. Actually, however, the whole matter might have been satisfactorily dealt with far more briefly—and it would probably have been more intelligible in consequence. For much of the formidable discussion on columns A to F does not really enter into the determination of the specific heats; nor do two at least of the columns given in the tables. The method that Wilcke finally adopted was simple enough, amounting to nothing more than the determination of the water equivalent of the body in terms of the mass of the body as unity. He takes a mass of ice-cold water equal to the mass of the body, heats the body to a known temperature (t), places it in the ice-cold water and observes the temperature of the mixture (θ). This finishes the experimental part of the determination. He then calculates, from Richmann's formula, what mass (m) of water at the initial temperature of the body would produce the same final temperature when mixed with the ice-cold water, the mass of the ice-cold water being unity, since it is equal to the mass of the body (which is taken as unity throughout). He now knows what mass of water, expressed as a fraction of the body's mass, produces exactly the same heating effect as the body, that is, the specific heat of the body. The calculation of this equivalent mass (m) of water in terms of the initial temperature (t) and final temperature (θ), either by Richmann's formula or by simple heat exchange, gives $m = \dfrac{\theta}{t - \theta}$ times the mass of the body, which is therefore the specific heat.

The only columns really necessary in the tables are column A, which gives the initial temperature (t) of the body; column B, which gives the temperature (θ) of the mixture; and column E, which gives the equivalent weight of water, or the specific heat $\left(\dfrac{\theta}{t - \theta}\right)$, or rather, the reciprocal of this number.

Column F is really a return to Wilcke's first attempt to determine the specific heat by measuring the quantity of snow (at $0°$) melted in reducing the temperature of the body from its initial temperature t to $0°$ C.—a method that embodies the principle of the ice calorimeter. Experimental difficulties had led Wilcke to discard a direct measurement of the mass of snow necessary, and instead to calculate it from his formula, substituting for the mass of the body the equivalent mass of water calculated from Richmann's formula. But, as Wilcke himself saw, this equivalent mass of water $\left(\dfrac{\theta}{t - \theta}\right)$ *is* the specific heat and the subsequent calculation of the mass of snow is unnecessary; in fact, to obtain the specific heat from the determined mass of snow, it is necessary to divide the mass of snow by $\dfrac{t}{72}$, the factor by which the water-equivalent has to be multiplied to obtain the mass of snow. The values given in column **F,** being the reciprocals of the mass of snow, must be multiplied, not divided, by this factor $\dfrac{t}{72}$ in order to obtain the specific heat from them. It is only for the particular value $t = 72$ that column F gives the specific heat itself, for then $\dfrac{t}{72} = 1$.

The purpose of columns C and D is to disprove earlier views on the distribution of heat. Column C gives the temperature that results when the body is replaced by an equal mass of water at the same temperature. The values obtained show that the quantity of heat in a body is not simply proportional to the quantity of matter in the body, that is, to its mass. If it were, a pound of water, gold, or lead, mixed with an equal mass of ice-cold water, would give the same temperature, and this temperature would be half the temperature of the body given in column A. This is obviously not the case; for, when unit mass of gold at $73°$ is mixed with unit mass of water at $0°$, the temperature is $3\frac{1}{2}°$, not $36\frac{1}{2}°$. Again, column D gives the temperature that results when the body is replaced by an equal volume of water at the same temperature, and shows at once that the heat is not distributed according to the volumes, as was held by Boerhaave (*ibid.*, p. 67).

We conclude the account of the experiments with the Table of results given by Wilcke (*ibid.*, p. 70). The denominators of the fractions in the second column are the mean values found in the experiments; their reciprocals are the specific heats. For convenience in comparison, the modern values of the specific heats are shown in brackets.

	Specific Gravity.	Specific Heat.
Water	$1 \cdot 000$	$1 \cdot 000 = 1 \cdot 000$
Gold	$19 \cdot 040$	$\dfrac{1}{19 \cdot 712} = 0 \cdot 050 \ (0 \cdot 0303)$
Lead	$11 \cdot 456$	$\dfrac{1}{23 \cdot 515} = 0 \cdot 042 \ (0 \cdot 0305)$

					Specific Gravity.	Specific Heat.
Silver	.	.	.	.	.	10·001	$\dfrac{1}{12·010} = 0·082$ (0·056)
Bismuth	.	.	.	.	.	9·861	$\dfrac{1}{23·082} = 0·043$ (0·0304)
Copper	.	.	.	.	.	8·784	$\dfrac{1}{8·750} = 0·114$ (0·0936)
Brass [1]	.	.	.	.	.	8·356	$\dfrac{1}{8·604} = 0·116$
Iron	.	.	.	.	.	7·876	$\dfrac{1}{7·933} = 0·126$ (0·119)
Tin	.	.	.	.	.	7·380	$\dfrac{1}{16·621} = 0·060$ (0·0552)
Zinc	.	.	.	.	.	7·154	$\dfrac{1}{9·730} = 0·102$ (0·093)
Antimony	.	.	.	.	.	6·107	$\dfrac{1}{15·818} = 0·063$ (0·0508)
Agate [1]	.	.	.	.	.	2·648	$\dfrac{1}{5·114} = 0·195$
Glass	.	.	.	.	.	2·386	$\dfrac{1}{5·326} = 0·187$ (0·16)

We will conclude our account of Wilcke's work with a translation of the passage in which he defines "Specific Heat," and which we passed over in dealing with his experiments. He says (*ibid.*, p. 54):

I now felt convinced, through experiments made with a great variety of different substances, that the quantity of heat [*Eld*, fire] in this distribution in different bodies and substances is not in general proportional either to the volume alone or to the density, or specific gravity, alone, but that each and every substance, according to its own particular nature, takes up, holds, and communicates a definite and fixed proportion of fire or heat-substance, the quantity of which, in comparison with that of another substance and in particular with water, can be called the body's *Specific Heat*, just as the term *specific*

---

[1] As the composition of the materials used by Wilcke is not known, we have been unable to give corresponding modern values.

weight is used to denote its weight in comparison with the weight of another body of the same volume. The specific heat, however, is in no way related to the specific weight, and so must be due to a property of the substance altogether different from the mere number of particles and pores in a given space, and their arrangement.

To avoid needless ambiguity and doubt in this comparison, it is necessary to distinguish between the *specific heat*, which applies to the substance itself, and that which is ascribed to the entire body as consisting of a definite volume of this substance. For when the question is asked: How much heat does one body contain in comparison with another? this comparison can be based either on the mass of the body or on its volume. In the first case, the question is: What is the relation between the quantities of heat in equal masses, or in equal numbers of material parts which are of different kinds, for example, between that in a pound of gold and that in a pound of water? and this is the same as the question how much heat does each and every separate particle in the body contain in comparison with that contained by a particle of another body, for example, water? The quantity of heat in this instance is called, in what follows, the Specific Heat of the substance of the body. Again, in the second case, the generally accepted meaning of the question is: What is the quantity of heat in a definite *volume* of the one body in comparison with the quantity of heat in the same volume of another body, for example, what is the quantity of heat in a cubic inch of gold in comparison with that in a cubic inch of water? Here the quantity of heat depends partly on the specific heat of the substance, partly on the density and number of the particles in the volume of the body. The specific heat in this instance is therefore defined, to distinguish it from the previous case, as the *relative heat* of the body as a whole. Both the specific heat of the material of the body and the relative heat of the body as a whole mean neither more nor less than the proportion of heat that each particle contains compared with another particle, or the proportion of heat that the body contains compared with another body, and thus both the specific and relative heats can remain unaltered, although the absolute or true quantity of heat increases or decreases by a definite number of so-called degrees. Thus the quantity of absolute heat ought to differ both from the specific heat of the particles or substance and from the relative heat of the body as a whole.

Before going on to describe Gadolin's work, it should be

pointed out that at this stage both Wilcke and Gadolin were acquainted with the work of Black, Crawford and Kirwan on specific heat. Wilcke, at the end of the memoir that we have just been discussing refers briefly to their researches (*ibid.*, p. 78), though his knowledge seems to have been obtained solely from a reading of Magellan's *Essai*. Black, as we have already shown, originated such investigations; and his experiments precede Wilcke's by twenty years (see pp. 31–3 above).[1]

Johan Gadolin (1784). The chief source for Gadolin's work on Specific Heat is a monograph that he published jointly with Nicolaus Maconi, *Dissertatio chemico-physica de Theoria Caloris Corporum Specifici* (Åbo, 1784). It is impossible to say how much Maconi contributed to this work; it is written throughout in the first person *singular*, and as Gadolin wrote two other memoirs dealing largely with specific heat, whereas Maconi seems to have made no independent contribution to the subject, we are probably justified in assuming that the real writer was Gadolin.

Gadolin begins by pointing out that when two equal masses of water at different temperatures are mixed, the temperature of the mixture is the mean of the separate temperatures. If the masses are not equal, then the temperature is given by

$$\frac{m_1 t_1 + m_2 t_2}{m_1 + m_2},$$

where m_1, m_2 are the masses, and t_1, t_2 the corresponding temperatures.

This is, of course, Richmann's Law.

If different substances are mixed, the temperature is not given by this formula; for the nature of the substance must then be taken into account. "The quantities of heat contained by bodies of different kinds but of the same mass differ from one another. To the ratio of these quantities is given the name

---

[1] It is interesting to find that in his later years Black referred in his *Lectures* (p. 79) to Wilcke's "valuable experiments" and advised his students to read the abstracts of Wilcke's papers given in the Leipzig Commentaries (*Comment de Rebus in Sci. Nat. et Med. Gest.*, Lipsiae, 1776, **21**, 633, and 1785, **26**, 197), evidently the only source of information known to him.

PLATE VI

JOHAN GADOLIN (1760–1852)

The famous Finnish chemist, Bergman's most distinguished pupil. Studied at
Åbo and Upsala. Friend of Scheele. Professor of Chemistry in the University
of Åbo (1797–1822). Discovered yttria, the first of the rare earths, in 1794,
the mineral from which it was obtained being subsequently called *gadolinite* after
him. When Marignac (1880) and Lecoq de Boisbaudran (1886) detected a new
element in samarskite, they named it *gadolinium* in honour of the discoverer of
the first of this curious group of substances, Gadolin thus becoming the only chemist
whose name is perpetuated in a chemical element. Gadolin carried out many
chemical researches: he visited England and worked with Crawford, and he met
Kirwan in Ireland. He adopted Lavoisier's theory in 1789 and wrote a Swedish
work on the new chemistry in 1798. He lived for thirty years after his retirement
and died at the age of 92.

(Reproduced by kind permission of Fru A. Gadolin, of Helsingfors.)

Specific Heat" (*op. cit.*, p. 5).   The specific heat of a substance is the same at all temperatures (*ibid.*, p. 6).

The quantity of heat in a body can be determined from the specific heat and the mass.   Thus if two bodies, whose masses are m_1, m_2, specific heats s_1, s_2, and temperatures t_1, t_2, are mixed together to give a temperature θ, it follows from the equality of the heats before and after, that

$$s_1 : s_2 = m_2(\theta - t_2) : m_1(t_1 - \theta).$$

This may readily be extended to mixtures consisting of several ingredients.   If all the specific heats except one are known, then that one can readily be found.   It is to be noted that only differences of temperature occur in the above expressions.

Though the determination of specific heats by mixtures in this way is very simple in theory, yet in practice several difficulties arise.   A correction must be made for the heat lost to the surrounding air or gained from it.   Allowance must also be made for the vessel; the greater the mass of the vessel, the more will it affect the result.   To reduce error due to this source, it is usually the practice to make the vessel of as thin metal as possible and to use comparatively large masses of the substance under examination.   As, however, the correction for the vessel is one of considerable importance, the matter calls for a separate investigation, and Gadolin accordingly proceeded as follows:

Let t_v be the temperature of the vessel under examination,

  m the mass of the liquid used,

  t the temperature of the liquid,

  θ the common temperature of the vessel and liquid when the liquid has been poured into the vessel,

  m' the mass of the same liquid which, at the temperature t_v of the vessel, produces with the given mass of liquid m the temperature θ of the mixture.

Then, since the quantities of heat, obtained by multiplying the masses by the corresponding temperatures, are equal before and after mixing, it follows that:

$$m' = \frac{m(t - \theta)}{\theta - t_v}.$$

The effect of the vessel on the temperature of the liquid is therefore equal to that of a quantity of the liquid equal in mass to

$$\frac{m(t - \theta)}{\theta - t_v}.$$

As the vessel may not always be filled to the same depth,
Gadolin examined the effect produced by taking it only one-
third, one-half, or three-quarters filled (*ibid.*, p. 9, footnote *d*).
He does not state what conclusions he reached. He adds that
his vessels were made of tinned iron; that they were very thin so
as to reduce the correction to a minimum; that they were raised
above the table on very slender supports so that little heat might
be lost through the base; and that in his experiments on saline
solutions, which readily attack metals, he used vessels made
of glass. The specific heats of the different substances are
expressed in terms of that of water taken as unity.

As the method used by Gadolin was not the same as that used
by others, he gives three examples to illustrate it. These are
given in a footnote extending over several pages (*ibid.*, p. 11,
footnote *e*).

Example I. *Lead*

6 loth [1] (m_2) of powdered lead was placed in a vessel whose
water equivalent was 0·6 loth (m') (*cujus vis alterandi calorem
injectorum tanta erat, quanta aquae lothonum* 0·6). The vessel
and powder were heated to 22·05° C. ($t_v = t_2$); 12 loth (m_1) of
water at 3·81° C. (t_1) was then added. Then

after 2 minutes the temperature was 6·46°,
   „ 3   „   „   „   „ 6·8°,
   „ 4   „   „   „   „ 7·15°,
   „ 5   „   „   „   „ 7·42°,
   „ 6   „   „   „   „ 7·7°,
   „ 7   „   „   „   „ 7·95°, and
   „ 8   „   „   „   „ 8·23°.

After the fourth minute the temperature increased steadily by
very nearly 0·27 degrees per minute. Gadolin therefore

---

[1] 1 loth = ½ ounce.

takes the best value for the temperature (θ) of the mixture as $7 \cdot 15 - 4(0 \cdot 27)$, that is, $6 \cdot 07°$. Remembering that the specific heat of the added water and the specific heat of the water-equivalent of the vessel are both 1, we have that s_l, the specific heat of the lead is given by

$$s_l = \frac{m_1(\theta - t_1) + m'(\theta - t_v)}{m_2(t_2 - \theta)},$$

that is, $$s_l = \frac{12(2 \cdot 26) - 0 \cdot 6(15 \cdot 98)}{6(15 \cdot 98)}$$

whence $s_l = 0 \cdot 183.$

This is, of course, a very poor value for the specific heat of lead, undoubtedly due to the unsatisfactory method used in arriving at the temperature of the mixture, which should be considerably lower than $6 \cdot 07°$ C. If we take $0 \cdot 03$ as the specific heat of lead, then we find that the temperature of Gadolin's mixture should have been slightly below $5°$ C. ($5°$ gives $0 \cdot 039$ for the specific heat, which is still too high.) The steady rise in temperature after the fourth minute would be due to the difference in temperature between the mixture and the surrounding air, the rise per minute tending to become less as the temperature of the air is approached. The rise ($0 \cdot 35$ degrees per minute) during the third and fourth minutes is mainly due to the greater difference between the temperature of the mixture and the air, though partly to the heating effect of the hot body, and the rise due to both these causes during the first and second minutes would be even greater. On the assumption that there is very little loss of time in the passage of the heat from the hot body to the cold, the temperature of the mixture should have been deduced from the rise of temperature during short intervals of time over the period immediately following the mixing, not from a reading taken four minutes after mixing and a rate of warming due (by that time) to the surrounding air alone. In any case, the result obtained could only be very uncertain, as a small error in the value assigned to the temperature of the mixture produces a serious error in the value found for the specific heat.

Gadolin considers that, in the investigation of specific heats, the process is simplest when the body is either a liquid or is in the form of a powder, as in the example just given. If a single piece of solid is used, the calculations are rather more complicated. In such a case the temperature of the body cannot be determined with any degree of accuracy unless the body is placed in hot water; but then some of the water adheres to the body when it is removed from the heater, and allowance must be made for this. Gadolin therefore determined the mass of water which thus adhered to the body, both for hot water and for cold (for the quantity was often rather greater for cold water than for hot). This correction is made in the two remaining examples.

EXAMPLE II.   *Copper*

Mass of copper (m_2) 21·4 loth. The copper was placed in warm water, removed, and weighed. The mass of the adhering water was found to be 0·07 loth (m_a).   15·25 loth (m_1) of water was placed in a vessel of water-equivalent 0·8 loth (m'). This was then heated to 16·3° $(t_1 = t_v)$. The copper was then transferred from the heater, temperature 39·8° (t_2, t_a), and suspended by a fine thread in the water. The temperature of the mixture was observed to be 19·06°, "and as this in a space of several minutes neither increased nor decreased, I took it to be the true temperature of the mixture" (*ibid.*, p. 12).

Remembering that the specific heat of the water, vessel (expressed as its water-equivalent), and adhering water is 1, the specific heat (s_c) of the copper is given by

$$s_c = \frac{m_1(\theta - t_1) + m'(\theta - t_v) + m_a(\theta - t_a)}{m_2(t_2 - \theta)},$$

whence $s_c = 0.097$.

This is very close to the modern value, namely, 0·0936. The value obtained by Wilcke was 0·114.

EXAMPLE III.   *A saturated solution of common salt*

This could not be mixed with water; for the mixing itself would result in a lowering of the temperature and so interfere with the result. The method adopted was therefore to immerse in the solution a body which would not produce this effect and whose specific heat was already known. Copper was chosen for this purpose, as its specific heat had been determined. In a preliminary experiment, the copper was immersed in a hot saline solution, removed and weighed; the mass of the adhering solution was 0·085 loth (m_a). 15 loth (m_1) of the solution was placed in a glass vessel of water-equivalent 1·2 loth (*cujus efficacia respondebat efficaciae aquae lothonum* 1·2). The temperature of the solution and vessel was observed to be 15·09° (t_1, t_v). The copper was removed from a similar solution at a temperature of 72·75° (t_2, t_a) and was suspended in the solution under test. After the fourth minute, when the temperature was 23·01°, the thermometer was observed to fall steadily at the rate of 0·043 degrees per minute. The temperature of the mixture (θ) was therefore taken to be 23·01 + 4(0·043) or 23·18°. The mass of the copper was 21·4 loth (m_2) and its specific heat (s_c) had been found to be 0·097.

Hence, since the specific heat of the vessel, expressed as its water-equivalent, is 1, the specific heat (s_s) of the solution is given by:

$$s_s = \frac{m_2 s_c (t_2 - \theta) - m'(t_v - \theta)}{m_1(\theta - t_1) + m_a(\theta - t_a)},$$

whence $s_s = 0·795$.

This concludes Gadolin's account of his experimental work. He next gives a table showing the specific heats of 91 substances, setting out in three separate columns the values that had been obtained by Kirwan and published by Magellan in the *Essai*, those obtained by Wilcke and published in the *Swedish Transactions*, and those that he himself obtained. With regard to his own values, he says that each is the mean of at least six, sometimes ten, experiments.

From an examination of this table Gadolin concludes that gases have the highest specific heat,[1] solids the lowest, and that the specific heat becomes less as the density increases. Inflam-

---

[1] See p. 43 above, footnote 4.

mable air, the lightest gas, has the highest specific heat, namely,
281·000 (!). When iron is made denser by beating, it loses
some of its specific heat.

If the specific heat depends on the degree of coherence of the
parts of the body, then it must be that a change in the state of the
body or of the union between its parts, is accompanied by a change
in the specific heat. This is very noticeable in the phenomena of
phlogisticated bodies. Inflammable air, such as is obtained from
metals by the aid of acids, consisting almost entirely of phlogiston
and without cohesion, has the form of an elastic fluid, the most tenuous
of all those examined, and is the richest in specific heat. Likewise
phlogiston loses to a remarkable degree its power of attracting heat
when it is present in either a solid or a liquid; or rather, it very often
diminishes by its union the specific heat of the other bodies with
which it is united. Thus acids rich in phlogiston have less heat
than those that are without it. Metals, which contain the inflammable principle in a very high degree, are inferior in their supply of
specific heat to all the calces, and to their own at least very nearly
in proportion to the loss of phlogiston.[1] Lead, which contains very
little phlogiston, has a specific heat only a little lower than its calx.
Tin and arsenic, which contain rather more phlogiston, show a greater
difference between the specific heats of the calx and metal (*ibid.*, p. 16).

The values given in the table for the specific heats mentioned
are:

Lead	0·050
Lead calx	0·068
Tin	0·068
Tin calx	0·096
Arsenic	0·084
Arsenic calx	0·126

It is known [wrote Gadolin] that snow and very many salts, when
they are dissolved in water, produce cold, and that liquids, when
they are vaporized, cause the thermometer to fall. The opposite
happens when the process is reversed: when water, cooled below the
freezing-point, solidifies, or a salt solution crystallizes, or a vapour

---

[1] Crawford had already carried out experiments proving this (*Animal
Heat*, 1779 edn., pp. 58–68).

condenses to a liquid, noticeable heat is produced. Such phenomena seem to indicate sufficiently clearly that a definite supply of latent heat is contained in all bodies (*caloris latentis quandam copiam omnibus contineri corporibus*) and that it is greatest in elastic fluids, least in solids. There is every justification for believing that the latent heat in bodies, which becomes sensible when a change of state occurs, in no way differs from the specific heat of the bodies. Iron, on being repeatedly struck, becomes hot and may be made to glow, and when it has been thus treated, its specific heat is smaller than before. This phenomenon also proves that there is the more heat hidden and bound (*latere ac ligatum quasi teneri*) in bodies that have the higher specific heat. There seems to be scarcely any room for doubt that sensible heat ought to arise when a change in the specific heat of a body occurs as a result of a change in its state or form (*ibid.*, p. 21).

Before leaving Gadolin, we must give some account, however brief, of his determination of the specific heat of ice. This work is described in his *Disquisitio de Theoria Caloris Corporum Specifici* (*Nova Acta Reg. Soc. Sci. Upsal.*, 1792, 5, 1).

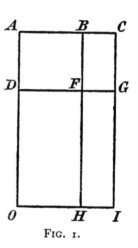

Fig. 1.

In order to follow Gadolin's work, it will be necessary to include the diagram on which his argument is to a large extent based. The use of such diagrams to illustrate the fundamental points of the theory of specific heat current at this time will be made clear if we give Gadolin's own example of their application to the determination of the "point of absolute cold." In Fig. 1 let AO represent the thermometric scale, O being the point of absolute cold and A any given temperature. Let AB be the specific heat of the body at the temperature A. Then, according to the theory, the total heat in the body is represented by the rectangle AH. Suppose that the specific heat for some reason increases to AC. Then, if the total heat of the body remains as before, the temperature must fall to some point D,

such that the rectangle DI equals the rectangle AH.   It
therefore follows that

$$AC.AD = BC.AO,$$

or $$AO = \frac{AC.AD}{BC}.$$

This gives the position of the absolute zero O below any stated
temperature A in terms of the specific heat at the temperature A,
the increase (or decrease) in the specific heat that occurs at some
other temperature D, and the position of D below A.   Gadolin
then gives (*loc. cit.*, p. 7) Irvine's determination of the position of
the absolute zero thus: according to Wilcke, snow in melting
absorbs as much heat as would raise the temperature of an equal
mass of water from 0° C. to 72° C., and the specific heat of
snow is $\frac{9}{10}$ that of water (Kirwan's value for ice, given in
Magellan's *Essai*).

Here, then AD = 72,

AC = 1, the specific heat of water, and

BC = 1 — 0·9, the difference between the specific
heats of water and snow.

Therefore $AO = \dfrac{72 \cdot \times 1}{1 - 0\cdot 9}$

$= 720.$

Gadolin leaves it at this and does not explain that the absolute
zero is 720 degrees below D, not A.   The rectangle ACGD
serves merely to determine the amount of heat required to convert snow at 0° into water at 0° by considering the equal
amount of heat required to raise the temperature of the same
mass of water from 0° to 72°.   In the case of the melting of the
snow, both A and D represent 0° C.

From Fig. 2 Gadolin deduces an expression for the heat
required to convert a quantity of snow [1] *below* 0° C. into water
at 0° C.   Here again AO denotes the thermometric scale, A

---

[1] Gadolin says that he made experiments with ice as well as with snow
and found no difference as regards absorption of heat.   He therefore used
snow in the experiments, as it could be procured more easily than ice (*ibid.*,
p. 12).

being the temperature of a given quantity of water, MO (= OP) the specific heat of the water (= 1).

Then rectangle AM represents the total heat of the water. Let OI be the absolute temperature of the snow, ON its specific heat. Then rectangle IN represents the total heat of the snow. D represents the common temperature of the water and snow when mixed. Then rectangle BD represents the heat absorbed by the snow, since it also represents the heat lost by the water.

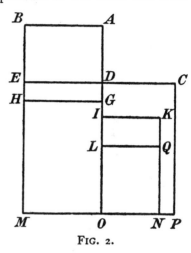

FIG. 2.

Let G be the temperature of the mixture when the same quantity of snow, but at a lower temperature L, is mixed with the same water as before. Then in this case the heat absorbed by the snow is represented by the rectangle BG. Then rectangle EG represents the difference in the quantities of heat required to raise the same quantities of snow at temperatures I and L to temperatures D and G respectively. Hence 2 rectangle EG represents the difference in the heat required in the two cases that the temperature of the snow may *in each case* become D. But this is the heat necessary to raise the temperature of the snow from L to I and is therefore represented by the rectangle IQ.

Hence rectangle IQ = 2 rectangle EG.

That is IL.ON = 2 DG.OP (since IK = OM, and ED = MO = OP).

Finally, rectangle BD = fig. IDCPNK, since the heat given up by the water in falling to D = heat absorbed by the snow in rising to D, the initial total heat of the snow being represented by rectangle IN,

that is, rectangle   BD = rectangle DP — rectangle IN.
$$= \text{rectangle IC} + \text{rectangle KP.}$$

∴   AD.DC = DC.DI + IO.NP, since AB = DC
and KN = IO.

∴   IO.NP = AD.OP — DI.OP,   since
DC = OP,

whence $$IO = \frac{OP(AD - DI)}{NP},$$

where OP is the specific heat of the water, i.e., 1,

        PN is the difference in the specific heats of the water and snow,

        AD is the difference in temperature between the water and the mixture, and

        DI is the difference in temperature between the snow and the mixture.

If the temperature of the water is such as to convert the snow (initially below $0°$ C.) into water at $0°$ C., then D represents $0°$ C. and AD is the temperature of the water say t [1], and DI is the temperature of the snow, say t_s.

Let s be the specific heat of snow, and $- T°$ C. the absolute zero.

Then OI = $T - t_s$,

and   PN = $1 - s$, since OP = 1.

Hence $T = t_s = \dfrac{1}{1 - s} \times (t - t_s)$,

whence $t = T - s(T - t_s)$ . . . . (1)

where t is the temperature of the water required to convert an equal mass of snow at t_s into water at $0°$ C.

The principle of Gadolin's method is to determine the value of t corresponding to different values of t_s; then to eliminate T from any two of the equations obtained by putting the values found for t and t_s in the equation

$$t = T - s(T - t_s).$$

The value of s, the specific heat of snow, is then known.

---

[1] We have here substituted modern symbols in place of the less familiar ones used by Gadolin.

To determine t, a mass m_s of snow at t_s is mixed with a mass m_w of water at a temperature t_w sufficiently high to melt all the snow.   Let θ be the temperature of the mixture.   Let m' be the water-equivalent of the vessel, its temperature being that of the water, t_w.

Then the heat given up by the hot water and vessel to the mixture is

$$m_w(t_w - \theta) + m'(t_w - \theta),$$

and the heat absorbed by the melted snow in rising from 0° C. to θ° C. is $m_s\theta$.

Hence the heat required to raise mass m_s of snow to 0° C. and to melt it without raising its temperature above 0° C. is

$$m_w(t_w - \theta) + m'(t_w - \theta) - m_s\theta,$$

or, for unit mass of snow,

$$\frac{m_w(t_w - \theta)}{m_s} + \frac{m'(t_w - \theta)}{m_s} - \theta.$$

This therefore must be the temperature of unit mass of hot water for it to convert an equal mass of snow at t_s into water at 0° C., i.e. it must be the temperature t given by expression (1). Hence

$$t = \frac{m_w(t_w - \theta)}{m_s} + \frac{m'(t_w - \theta)}{m_s} - \theta \quad . \quad . \quad (2).$$

Though both expressions (1) and (2) refer to *equal* masses of water and snow (a necessary restriction resting on the definition of specific heat), yet the masses of water and snow used in the experimental determination of t, as expressed by (2), need not be equal.

A single example from Gadolin's work will be sufficient to show the method of procedure.   In his first experiment (*ibid.*, p. 14), he found:

Mass of the snow (m_s) = 8 loth.
Temperature of the snow (t_s) = 0·08° C.
Mass of water (m_w) = 32 loth.
Temperature of the water (t_w) = 30·96° C.
Water-equivalent of the vessel (m') = 0·66 loth.
Observed temperature of the mixture (θ) = 8·38° C.

Substitution of these values in (2) gives $t = 81.242$.   As the mean of nine such experiments, in which t_s was always between $0°$ C. and $-1°$ C., Gadolin obtained the value 81.452.   The mean value of t_s was $-0.38°$ C.   He then carried out a second series with t_s always between $-1°$ and $-2°$ C.   The mean value for t in this series was 81.947, the mean value of t_s being $1.4°$ C.   In all, he carried out 15 such series of experiments, t_s in the last series being between $-15°$ and $-21°$ C.

The next step is to eliminate T from expression (1) by using any two pairs of mean values of t and t_s.

Let t, t' be any two values of t,

$\quad t_s$, t_s' any two values of t_s.

Then on eliminating T from the two equations obtained by putting these values in (1), we find:

$$t - s.t_s = t' - s.t_s',$$

whence $$s = \frac{t - t'}{t_s - t_s'}.$$

Thus from the means of the first two series of experiments given above, we have

$$s = \frac{81.947 - 81.452}{1.4 - 0.38}$$
$$= 0.485.$$

The mean values of all the 15 series of experiments are taken together in pairs in this way, thus giving no less than 105 values for s, all of which agree very well among themselves.   Finally, Gadolin took the mean of all these 105 values, obtaining 0.52 (actually stated as 0.52408) as the specific heat of snow, remarkably close to the value now accepted, namely, 0.502.

By substituting this value for s in (1), Gadolin found, as the mean of the 15 series of experiments, $T = -170.6°$ C.

The consistency of the results obtained by Gadolin is well brought out in the following table, with which he concludes this series of experiments:

t_s	$T - s(T - t_s)$	$T(1 - s)$	T
0·38	81·452	81·25	170·7
1·4	81·947	81·21	170·6
2·48	82·614	81·31	170·8
3·59	83·215	81·33	170·9
4·47	83·320	80·98	170·2
5·48	84·040	81·17	170·6
6·62	84·315	80·85	169·9
7·66	85·248	81·23	170·7
8·42	85·647	81·23	170·7
9·48	86·084	81·12	170·4
10·55	86·760	81·23	170·7
11·44	86·910	80·92	170·0
12·53	87·783	81·22	170·7
14·05	88·724	81·36	171·0
17·45	90·571	81·43	171·1
		81·19	170·6

The values given in the column headed $T(1 - s)$ are actually values for the latent heat of fusion of snow.[1]   But they serve merely as a step in calculating T; for, beyond giving their mean, Gadolin makes no reference to them whatever.

[1] Since t is the temperature of the water that will convert an equal mass of snow at t_s into water at 0°, and st_s is the heat absorbed by snow when its temperature rises from t_s to 0°, therefore $t - st_s$ represents the heat necessary to convert snow at 0° into water at 0°.   But, from $t = T - s(T - t_s)$, it follows that $t - st_s = T(1 - s)$.

CHAPTER V

SOME IMMEDIATE CONSEQUENCES OF THESE DISCOVERIES

In discussing some of the immediate consequences of the discoveries of specific and latent heats, we do not propose with regard to Watt's improvement of the steam-engine to do any more than point out that his invention of the separate condenser, probably the most important technical advance of the nineteenth century and one that may be truly said to have ushered in a new era of industrial development, was directly derived from Black's discovery of latent heat: we might add that Wilcke's studies had nothing at all to do with this, nor could they, because Wilcke's earliest work in this field was not carried out until 1772 and more especially because he never investigated the latent heat of vaporization of water or of any other substance. In other directions, however, there were several advances, practical and theoretical, that deserve our consideration. These include the rise of calorimetry or, in other words, the establishment of the quantitative science of heat, the various attempts to determine the zero of temperature or, as it was called at that time, "the point of absolute privation of heat," and the new significance that could be given to experiments devised to weigh the so-called "matter of heat."

The Rise and Progress of Calorimetry. The most important result of Black's work was the rise of calorimetry, the beginnings of the quantitative science of heat: for Black had devised a method of measuring heat exchanges in terms of a temperature change of so many "degrees of heat on Fahrenheit's scale" for an equal mass of water, and, although this is not the method followed nowadays, it was a thoroughly satisfactory working instrument at that time.   Heat had thus become measurable in terms of its temperature effects, and the quantitative study of thermal phenomena made rapid and extensive progress once this first step was taken.

Black, as we have already pointed out, did not describe the method that he employed in determining what he called the "capacity for heat" of a substance.   According to Robison, he

estimated the capacities by mixing the two bodies in equal masses, but of different temperatures; and then stated *their capacities as inversely proportional to the changes of temperature of each by the mixture*. Thus, a pound of gold, of the temperature 150°, being suddenly mixed with a pound of water, of the temperature 50°, raises it to 55° nearly: Therefore the capacity of gold is to that of an equal weight of water as 5 to 95, or as 1 to 19; for the gold loses 95°, and the water gains 5° (p. 506).

We have also Robison's statement:

Before the year 1765, Dr. Black had made many experiments on the heats communicated to water by different solid bodies, and had completely established their regular and steady differences in this respect.   He mentioned these in his lectures in the University of Glasgow, and was occupied with the experiments for the same purposes when he was called to Edinburgh.   He was much assisted in the experiments by Mr. Irvine, who afterwards lectured from the same chair.   The registers of these experiments remain among Dr. Black's papers, most of them dated, and partly written by himself, and partly by Mr. Irvine.   They are all previous to 1770 (p. 504).

The only other information we have on this point is from Crawford, who says that Black and Irvine found "the heat of iron . . . to be to that of water, as one to eight" (*Animal Heat*, 1779 edn., p. 62), adding in a footnote in the same place but without giving values that they had also determined "the heat of lead and tin" previous to his own experiments.

Robison in his note went on to state that it was

most convenient to compare all bodies with water, and to express the capacity of water by unity, or to call it 1.   Let the quantity of the water be W, and its temperature w.   Let the quantity of the other body be B, and its temperature b.   Let the temperature of the mixture be m.   The capacity of B is $\dfrac{W \times \overline{m-w}\,^1}{B \times \overline{b-m}}$ Or, when the water

has been the hotter of the two, the capacity of B is $\dfrac{W \times \overline{w-m}}{B \times \overline{m-b}}$.

In other words, *multiply the weight of the water by its change of temperature.   Do the same for the other substance.   Divide the first product by the second.   The quotient is the capacity of the other substance, that of water being accounted 1"* (p. 506).

Omitting for the moment any discussion of the taking of water as the standard substance, we note that Robison does not ascribe this calculation to Black, nor indeed to anyone else, but it appears that the formulation of this rule, applicable in cases where unequal quantities were involved, was due to Irvine, since it is ascribed to him by Crawford (*Animal Heat*, 1779 edn., p. 37; 1788 edn., pp. 121 and 235) and also by his (Irvine's) son, William Irvine, who, urged by Robison to search among his father's papers, published with certain specified contributions of his own what he could find among these papers bearing on the researches in which his father had been engaged (*Essays, chiefly on Chemical Subjects*, London, 1805, p. 86).   It should be noted that Irvine's papers were far from complete and that his son had great difficulty in putting together the disjointed material that composed them.   But it appears from Crawford's statement that Irvine had established this rule at least as early as 1779: it was probably derived much earlier than this (see Robison's note quoted above).

What is of more interest is Robison's next comment:

These experiments require the most scrupulous attention to many circumstances which may affect the result.   1. The mixture must be made in a very extended surface, that it may quickly attain the medium temperature.   2. The stuff which is poured into the other should have the temperature of the room, that no change may happen

---

[1] In modern notation, $s = \dfrac{w(\theta - t_1)}{m(t_2 - \theta)}$, or, when the water is the hotter,

i.e. $t_1 > t_2$, $s = \dfrac{w(t_1 - \theta)}{m(\theta - t_2)}$.

in the pouring it out of its containing vessel. 3. The effect of the vessel in which the mixture is made must be considered. 4. Less chance of error will be incurred when the substances are of equal bulk. 5. The change of temperature of the mixture, during a few successive moments, must be observed, in order to obtain the real temperature at the beginning. 6. No substances should be mixed which produce any change of temperature by their chemical action, or which change their temperature, if mixed when of the same temperature. 7. Each substance must be compared in a variety of temperatures, lest the ratio of the capacities should be different in different temperatures. When all these circumstances have been duly attended to, we obtain the measure of the *capacities* of different substances for heat (p. 506).

Here, of course, Robison is writing of the method in use at the time when he was editing the *Lectures*, namely, in 1803. As far as possible, we shall endeavour to trace these various improvements in calorimetric method to those who should properly receive the credit for their introduction.

In the first place, it is necessary to point out that Black had already, in 1760, allowed for the thermal capacity of vessels in experiments of another kind, this procedure incidentally providing further interesting evidence of the early date of his ideas about specific heat, or rather the "capacity" of bodies for heat. Black, unaware that the problem had been investigated previously by Brook Taylor (*Phil. Trans.*, 1723, 32, 291) and others indeed before Brook Taylor, questioned whether the expansion of the liquid in a thermometer was proportional to the increment of heat. He accordingly tested a thermometer by taking its readings when it was immersed, in the first place, in one part of cold water and then in a series of mixtures of 1 part of cold water with 1, 2, etc. parts of hot water, the result being communicated to the Glasgow Philosophical Society (pp. 57–8). In describing this test, he said that it was necessary

to make allowance for the heating or cooling of the vessel in which the mixture is made.[1] If the warm water is passed into the cold,

---

[1] This had not been taken into account by Brook Taylor or by any previous experimenter.

the cold vessel in which the mixture is thus made, will take from the mixture some of its heat, and make it appear colder than it should otherwise do. If, on the contrary, the cold water is poured into the hot, the hot vessel will impart some of its excess of heat to the more temperate mixture, and thus raise its heat above the degree which should be produced. To avoid this deception, we must employ the two vessels of the same materials, and of the same size and weight; and then by making the experiment both the one way and the other, and taking the medium of the results, we shall learn the truth (p. 58).

Further, correction for the thermal capacity of the containing vessel had likewise been applied by Black in 1762 in his determination of the latent heat of fusion of ice by the method of mixture, where he ascertained by other experiments that the "capacity for heat" of 16 parts (by weight) of hot glass was the same as that of 8 parts of equally hot water and substituted in his calculation 8 half-drachms of warm water in place of 16 half-drachms of warm glass, which brought the effective quantity of warm water up to 143 half-drachms (see pp. 18–19 above). Crawford followed this method (1779 edn., pp. 21–2; 1788 edn., pp. 99–100). The correction for the thermal capacity of the vessel was an obvious one to the discoverer of specific heat, and he accordingly made it; but, as we have already seen (pp. 85–7 above), Wilcke in his experiments of 1772 on the melting of snow omitted this important correction, clearly because at that time he was totally unaware of the idea of "specific heat," with the result that his value of "72 degrees" was considerably lower than what he would have otherwise obtained, if he had made the necessary allowance for the vessels that he used in his determinations.

The fifth caution advised by Robison is, in effect, a caution to allow for the heat lost by the system to the surroundings in the first few moments after the substances have been mixed, and, if he had been more explicit, he would have set it down as the application of Newton's Law of Cooling. It appears that Irvine was responsible for this refinement; for Crawford, in describing his own experimental method, states: "I must here observe, that I believe Dr. Irvine was the first who applied this rule of Sir Isaac Newton, to calculate the heat lost during the first minute, on experiments for determining the absolute heat of bodies" (*Animal Heat*, 1779 edn., p. 21). Crawford repeated this in the second edition of his book (1788, p. 99). And we should here observe that, previous to publishing his second edition, Crawford wrote to Irvine with regard to the insinuations then current, that he (Crawford) had without acknowledgment published part of Irvine's discoveries in the first edition, and received a reply in a letter from Irvine inscribed "Glasgow College, Jan. 27, 1780," to the effect that "there is nothing in your publication . . . that I lay any claim to, which you have not in that work already ascribed to me" (Crawford, *Animal Heat*, 1788 edn., pp. 484–5). William Irvine, the younger, also ascribed this to his father (*Essays*, pp. 28–30), but strangely enough gave Crawford's book as his authority, although independent proof was possibly not forthcoming from Irvine's papers on account of their incompleteness. But it is evident that this improvement is ascribable to Irvine, and, although we cannot say when he introduced it, it is clear that it was already made when Crawford began his experiments at Glasgow in 1777. As an instance, the first we are aware of, we may quote an experiment from Crawford's *Animal Heat* (1779 edn., pp. 21–2):

Temperature of air in room . . . . 61° F.
A pound of water at 168° F. was poured into an earthen vessel at 68° F.
The temperature of the water at the end of

1 minute was . . . . . 155° F.
2 minutes . . . . . . 150° F.
3　　" . . . . . . 145° F.

Crawford then calculated thus: "To discover the heat communicated to the atmosphere in the first minute, say as 84 to 89 so is 94 to a fourth proportional, which gives 99·5. From which it appears that 5·5 were carried off by the air in the first minute, adding 5·5 to 155 we have 160·5 for the true temperature of the water and of the vessel."

Crawford noted however that, when experiments were

carried out in vessels that transmitted heat very slowly and when the temperature of the mixture was little in excess of that of the atmosphere, the amounts of heat lost in small successive intervals of time were practically constant and correction could be made in a much simpler way (*ibid.*, p. 20). For example, he had himself found that the temperature of a pound of water at 120° in an earthenware vessel fell during 7 successive minutes to 119°, 118°, 117° . . . 113°: and the method of correction in such conditions was obvious.

The other precautions advised by Robison are of a more general kind and we cannot say who was responsible for their introduction.

Before leaving our account of these developments in calorimetric method, we must note the introduction of a method which seems to be due to Crawford and which was described in the second edition of his *Animal Heat*. Here Crawford wrote:

It was before observed, that the mixture will cool slowly if its temperature do not much exceed that of the external air. It may be proper to add, that if the colder substance be taken at a lower temperature than the air in the room, and the warmer substance at a higher; and if the difference of their heats be so proportioned, that the common temperature, after they are mixed together, shall be nearly the same with that of the air in the room; the fallacy, arising from the refrigeration of the mixture, will be altogether avoided, at the same instant that a large scale may be obtained (p. 108).

As far as we are aware, this is the first mention of such a method: and, since it is not described in Crawford's 1779 edition and since we cannot trace it in contemporary works, it seems probable that it was devised by Crawford himself at some date between 1779 and 1788.

It is curious that it is not at once clear who introduced the practice of taking water as the standard substance, that is, of taking the specific heat of water as being unity and comparing the specific heats of all other substances with that of water. Black, as we have already pointed out, determined the "capacity for heat" of gold compared with that of an equal mass of water as 1 to 19, and, similarly that of iron to that of water as 1 to 8, which is the opposite of taking water as unity, although in some sense it does take water as the standard substance of reference. At any rate, Black compared equal masses, where, as we have seen, Boerhaave had studied equal bulks and thereby greatly confused himself. Crawford, however, in the first edition (1779) of his *Animal Heat* (p. 22) wrote that in this work "water is generally the standard." An examination of the results that he gave for his determinations of "the comparative quantities of absolute heat in bodies" (i.e. specific heats) shows that he followed the same practice as Black and also as Irvine, writing, for example, the "comparative heat" of water to that of wheat as 2·9 to 1, and that of tin to that of water as 1 to 14·7. He usually compared substances with water, except when his particular purposes demanded that a substance should be compared with another substance that was not water, but water is not his standard in the sense of its "comparative heat" being taken as unity and he drew up no table of "comparative heats" in the first edition of his *Animal Heat*. He did, however, give the specific heats of two gases in terms of that of water taken as unity.

In the second edition of his *Animal Heat*, published in 1788, Crawford followed the same practice in working out the results of his experiments, except that as before in calculating the "comparative heats" of gases he took water as unity. In the tables drawn up by Crawford in various places in this work, the "comparative heats" of all the substances with which he had experimented were, however, recalculated and referred to that of water as being equal to unity (pp. 139, 287, 300 and 489). This edition was dedicated to Kirwan,[1] who had been present when some of the experiments were carried out (p. 204); and, as we have already seen, Magellan in his *Essai* (1780) gave a table, the first published table, of specific heats which had been communicated to him by Kirwan and in which the specific heat of water was taken as unity. We have also seen that Wilcke took

[1] The first edition was dedicated to John Watkinson.

I

water as the standard substance, setting down its specific heat as unity, and so did Lavoisier and Laplace; and we have shown that both Wilcke and Lavoisier and Laplace were familiar with Magellan's *Essai* and hence through that with Kirwan's system. It therefore appears that Kirwan must be credited not only with drawing up the first table of specific heats, but also with introducing the practice of taking water as the standard substance of reference with a specific heat of 1.

It is curious to observe that Kirwan wrote "specific fire" rather than "specific heat," with one exception to which we shall refer presently: and it is regrettable that we know his work in this field only through the medium of Magellan's *Essai*. In various places Kirwan refers to "specific fire" (*Phil. Trans.*, 1782, **72**, 196, 197, and 201; *ibid.*, 1784, **74**, 167 and 168) and elsewhere he wrote: "All bodies require a certain quantity of elementary fire or light to heat them to a certain degree, but the quantity requisite to produce this degree varies, according to the nature and species of these bodies, and hence the proportion suited to each is called their *specific* fire" (*An Estimate of the Temperature of Different Latitudes*, London, 1787, p. 38). On one occasion he used both terms, "specific heat" and "specific fire" (*Essay on Phlogiston and the Constitution of Acids*, London, 1787, p. 26). It thus seems that Magellan's *chaleur spécifique* marks the first appearance of the term "specific heat," although as we have already seen it was not there used in exactly the same sense in which it is used nowadays.

Attempts to Determine the Zero of Temperature. As a result of Black's discoveries various attempts were made to determine the zero of temperature or, as it was generally called at that time, "the point of total privation of heat." These efforts were described by Black as ingenious, but not satisfactory (p. 64). It appears that the method on which they depended originated with Irvine. We have already pointed out that Irvine did not publish his work; and we have to rely on the accounts given by his contemporaries for our knowledge of his investigations and on his posthumous *Essays*, already referred to in the preceding section. It is from these *Essays* that we shall describe Irvine's work, while his priority will be attested from the writings of his contemporaries.

Irvine (*Essays*, pp. 116–17) studied the case of a body capable of assuming both solid and fluid states and argued that, if the capacity of the body in the solid state is to its capacity in the fluid state as 1 to 2, then there must be twice as much heat in the body when it is fluid as there is in it when it is solid.   Therefore, if the total heat in the solid is 100 "degrees," 100 "degrees" become latent when the body passes from the solid to the fluid state.   Similarly, if these capacities are as 1 to 3, the total heat in the solid being 100 "degrees" as before, then the total heat in the body when fluid is 300 and the latent heat 200 "degrees." Further, if the capacities are as 10 to 11, the total heat in the body when fluid will be 110 and its latent heat 10 "degrees." The rule that Irvine induced from the foregoing will be clearer if we first tabulate his data thus:

Total heat in solid = 100 "degrees"

Ratio of capacity of solid to that of fluid.	Total heat in fluid.	Latent Heat.
1 : 2	200	100
1 : 3	300	200
10 : 11	110	10

Irvine then proceeds: "And universally the whole heat in the solid deing 100, the latent heat in the fluid will be equal to the difference of the capacities of the solid and fluid, divided by the number expressing the capacity of the solid, and multiplied by the whole heat of the solid."   For an example, he quotes thus: "When the capacities are as 10 : 11, then the difference 1 divided by 10, the capacity of the solid quotes $\frac{1}{10}$, which multiplied by 100, the whole heat of the solid at the melting-point, gives 10 for the latent heat in this instance" (*Essays*, pp. 117–18).[1]

From this Irvine then went on to show how, if the latent heat was given and the capacities of the solid and fluid forms known,

---

[1] This was one of the very few passages that the younger Irvine was able to quote from his father's disordered and incomplete papers.

the total quantity of heat contained in the body might be
calculated.   As before, if the total heat in the body when solid
was 100 "degrees" and if the capacities of solid to fluid were as
1 to 2, the latent heat was equal to the total heat contained in the
body when solid; if the capacities were as 1 to 3, the total heat in
the solid was equal to a half of the latent heat; and, if they were
as 10 to 11, the total heat in the solid was equal to ten times the
latent heat.   Again, it will be clearer if we tabulate the foregoing before recording the rule that Irvine derived thence.

Latent Heat $= l$

Ratio of the capacity of solid to that of fluid.	Total heat in solid.
1 : 2	l
1 : 3	$l/2$
10 : 11	$10l$

The rule, as given by Irvine the younger, was as follows:

And universally, the whole heat will be equal to the capacity of
the solid multiplied by the latent heat, and divided by the difference
of the capacities.   Or, in other words, the difference of the numbers
expressing the capacities in the solid and fluid form, is to the number
expressing the capacity of the solid, as the latent heat is to the absolute heat of the solid at the melting-point. . . .   These investigations
may be expressed by means of algebraical symbols.   Let $x =$ the
absolute heat of the solid body expressed in degrees: Let a and b
represent the capacities of the fluid and solid forms, and $l =$ the
latent heat.   Then it was found that $l = \dfrac{xa - xb}{b}$; or making

$a - b = c$, then $l = \dfrac{cx}{b}$, and $x = \dfrac{bl}{c}$.   Or in numbers, if we suppose the capacity of ice to that of water to be as 9 to 10, and the
latent heat to be 140, then $\dfrac{1 \times 140}{1}$ [1] $=$ the absolute heat of ice at
$32 = 1260$: if we suppose the capacities to be as 8 to 9, the absolute

---

[1] Given thus in original: it should obviously read $\dfrac{.9 \times 140}{.1}$

heat will be = 1120; if as 7 to 8, then 980, if as 6 to 7, then 840 (*Essays*, pp. 121–2).

We must note, firstly, that the latter quotation is from Irvine the younger, while the former was taken from the papers of Irvine himself and included by Irvine the younger in the *Essays*; and, secondly that determination of what was called "the absolute heat of the solid body expressed in degrees" was equivalent to determination of the "point of total privation of heat," i.e. the zero of temperature.

It is interesting too to note, in connection with Irvine's view of "capacities" and of the heat that Black called latent, that, as we have already seen, Irvine considered that heat entered a melting body because at the melting-point the capacity of the solid body was increased; and that Irvine the younger stated that his father was thus "induced to conceive that it was possible that the capacity of water might be found to exceed that of ice" and was thence led to experiment (*Essays*, p. 52).[1]

With regard to the method used by Irvine for the determination of the capacity of ice, Irvine the younger informs us as follows:

He found the capacities of some suitable bodies, as river sand, or iron filings, and compared them with that of water in the usual manner. This being done, he used the same body to examine the capacity of pounded ice formed from distilled water, or of snow. The temperature of the room and vessel was, in his experiments, always either 32° or below it; most commonly considerably under 32°.[2] He then took a known weight of snow or ice of a known temperature, in a vessel of which the capacity was determined by experiment. Upon this he poured a certain quantity of river sand washed, or iron filings of a certain temperature, with as much rapidity as possible: the new temperature was observed after stirring, and allowance was made for the heat gained or lost: the temperature of the mixture was frequently 19°, 20°, 25°, 16°,[2] etc. So that in a room where the air was below the freezing-point, the accuracy of the result could not be affected by the formation of any water; still there are many sources

---

[1] Crawford makes the same statement (*Animal Heat*, 1788 edn., pp. 483–4).
[2] Fahrenheit.

of inaccuracy remaining. But in Dr. Irvine's hands, the capacity of ice always turned out to be less than that of water. In all his experiments, which were very numerous, and repeated with care for many succeeding years, he arrived at results approximating to each other, and concluded, to use his own words, that from the mean of all his trials, the capacity or relative heat of water to that of ice is not in a ratio greater than 5 to 4 or 10 to 8 (*Essays*, pp. 55–6).

The same figure is given elsewhere (*ibid.*, p. 135).

Accordingly, by inserting this value in the formula already given, it will be found that the zero of temperature is at $\dfrac{8 \times 140}{2}$, or 560 degrees below freezing-point, i.e. at $- 528°$ F. This value is nowhere given by Irvine. It appears that he obtained a value of $- 900°$ F., which he did not regard as conclusive and about the determination of which no details are available. But it appears that he obtained this value by the method we have just described and also by measuring the heat developed on mixing sulphuric acid and water (*Essays*, pp. 127 and 137), such heats as the latter, now recognized as heats of chemical reaction, being then and subsequently regarded as latent heats. However, if we insert in the formula $l = (xa - xb)/b$, the values of $l = 140$ and $x = 900 + 32$, we obtain for the ratio of a to b a value of 10 to 8·7,[1] so that to have found a figure of $- 900°$ F. for the zero of temperature, Irvine must have used a value of 10/8·7 for the ratio of the capacities of water and ice, although no such value is mentioned anywhere in the accounts of his work, either in the posthumous *Essays* or elsewhere, his only assertion on this point being that the ratio never exceeded 10 to 8.

With regard to Irvine's priority here, we find that no mention of this work appears in the first edition (1779) of Crawford's *Animal Heat*, but that the first published account of work of this

---

[1] $l = (xa - xb)/b$, i.e. $140 = 932(a - b)/b$.
∴ $140b = 932a - 932b$, or $1072b = 932a$,
    whence $a/b = 10/8·7$ (nearly).
Putting $a/b = 10/8·7$ and $l = 140$ in the formula, the value of x becomes $- 937$ degrees of Fahrenheit's scale, whence the zero of temperature is at $- 937 + 32$, or $- 900°$ F. approximately.

kind appears in Magellan's *Essai* (p. 176), where it is ascribed to Kirwan and where the specific heat of ice is taken as 0·9, a value which Magellan ascribed to Kirwan and which is there used to determine what Magellan called the "absolute quantity of specific heat," i.e. the total heat in ice at the melting-point, the value obtained being 1166·4 [1] "degrees" (Fahrenheit), since Magellan used Wilcke's figure of 72 "degrees" on the Swedish scale for the latent heat of fusion of ice, taking it as equal to 129·6 degrees of Fahrenheit's scale (*Essai*, pp. 172–3).

Crawford, in the second edition (1788) of his *Animal Heat*, ascribed the idea to Irvine (pp. 265 and 483–4). Meanwhile Nicholson, who gave no account of it in the first edition (1782) of his *Introduction to Natural Philosophy*, ascribed it in the second edition (1787) to Kirwan (Vol. II, p. 119, footnote), correcting this however in the third edition (1790) and later editions by ascribing it to Irvine (3rd edn., Vol. II, p. 119, footnote). We have already seen (p. 32 above) that Robison ascribed it to Irvine and that Irvine's son claimed it for his father (*Essays*, p. 115).

Gadolin, too, worked on this problem. Taking Black's value of 147 "degrees" of Fahrenheit's scale for the latent heat of fusion of ice as equivalent to 81⅔ "degrees" of the Centigrade thermometer and also Wilcke's figure of 72⅙ "degrees" for the heat absorbed in the melting of snow, he applied the value of $\frac{9}{10}$ for the specific heat of snow (taking this value from Magellan's *Essai*, where it was given as the specific heat of ice) and calculated that the zero of temperature was at $-817°$ C. according to Black's and at $-722°$ C. according to Wilcke's figure (*K. Svenska Vet. Akad. Nya Handl.*, 1784, 5, 218). Gadolin's own experiments are described summarily, and it appears that he found the absolute zero to be at $-800·6°$ C. (a mean value of several determinations) by a method practically identical with that which he used later on and which we have already discussed at the end of the last chapter (see pp. 115–21 above).

Gadolin also, like most of Irvine's imitators, applied the latter's theory to heat changes other than those occurring in

[1] $x = (9 \times 129·6)/1 = 1166·4.$

liquefaction, for example, to the heat changes occurring in the solution of common salt in water and in the mixture of oil of vitriol with water, and found that the experimental values obtained were in agreement with the previous figure, namely, —800° C. (see also Crawford, *Animal Heat*, 1788 edn., pp. 457–77).[1]

Crawford, from a study of the combustion of "pure" and "inflammable" airs, i.e. oxygen and hydrogen, found the zero of temperature to be at —1551° F. (*Animal Heat*, 1788 edn., pp. 263–5). Incidentally, we find that the younger Irvine (*Essays*, p. 55) quotes a value for the ratio of the capacity of water to that of ice as 10 to 9 and ascribes this to Crawford, although there is no mention of this determination in Crawford's *Animal Heat* nor is ice included among the substances in his tables of "comparative heats" referred to above.

Meanwhile, Lavoisier and Laplace (*loc. cit.*) also applied Irvine's theory and, using their experimentally determined figure of 60 "degrees" on Reaumur's scale for the latent heat of fusion of ice, obtained the values given below for the zero of temperature:

Mixture.	Number of degrees below zero on Reaumur's scale.
Water and quicklime in the proportion of 9 to 16 . .	1537·8
Water and oil of vitriol in the proportion of 3 to 4 .	3241·9
Water and oil of vitriol in the proportion of 5 to 4 .	1169·1
Nitric acid and quicklime in the proportion of $9\frac{1}{3}$ to 1 .	1889
	—0·01783

The last value, as they pointed out, was physically impossible. Using Kirwan's figure of 9/10 for the specific heat of ice, they obtained a value of — 600° R. for the zero of temperature. The wide discrepancy between the values obtained, they argued, cast doubt upon the method by means of which they had been derived and, moreover, it was not established beyond doubt that the increments of heat in bodies for equal increments of tem-

---

[1] Crawford's account is misleading. Gadolin did not determine the latent heat of fusion of snow, as stated by Crawford, but proceeded as indicated above.

perature (i.e. their specific heats) were proportional to their
absolute heats (i.e. the total heat in the bodies). However,
Lavoisier and Laplace admitted that a slight alteration, of an
order not exceeding one-fortieth, in the values of the specific
heats obtained would be sufficient to make all the values agree,
and they accordingly recognized that their results were neither
for nor against the theory and all that could be concluded from
them was that, if the theory was true, the zero of temperature
was at least 600 degrees below freezing-point on the Reaumur
scale, since, to reduce this value, they would have to admit a
greater inaccuracy in their experimental method than was
involved in it.

Further doubt was cast on this method by Gadolin (*Nova
Acta Reg. Soc. Sci. Upsal.*, 1792, 5, 1), who from Wilcke's value
of 72 "degrees" for the latent heat of fusion of snow and Kirwan's value of 9/10 for the specific heat of ice calculated the
zero of temperature as − 720° C., and from his own experiments described above (pp. 115–21) found the zero of temperature to be at − 170° C., while similar determinations with wax
indicated that the zero was at − 480° C. Referring to these
and to Crawford's value of − 1532° F. or −851° C., Gadolin
concluded that the method was unsatisfactory and that the
specific heats of bodies were neither constant with temperature
nor proportional to the total heats of bodies: and after the
appearance of this memoir no further serious interest seems
to have been taken in the matter.

The " Weight of Heat." Throughout the eighteenth
century heat was widely regarded as a variety of matter; and it
was a commonplace among the philosophers of that time to
speak of the "matter of heat." Yet, as we have already seen,
the great advances made by Black were independent of any
theoretical speculations about the nature of heat. Indeed,
Black's Newtonian approach to scientific problems thrust aside
all ultimate hypotheses as mere waste of so much time and
ingenuity that might be more profitably employed in other ways.
And, while in the previous century much had been done both
by Hooke and by Boyle towards the establishment of a dynamical

theory of heat, of a theory that heat was a form of motion, it was Boyle's own work on another problem that contained the germ of the material theory of heat that in the eighteenth century finally dominated scientific thought in this field: for in his *Essays of Effluviums* (London, 1673) Boyle had argued from extensive experiments that the gain in weight shown by metals on calcination was due to the fixation in them of the material particles of fire.[1] To trace the growth of this and the opposing view is not our purpose here: and it will suffice to refer briefly to a few of the more important experiments precedent to the discoveries with which we have been concerned above and to show how experiments of this kind acquired a new significance through Black's investigations.

First among the numerous workers on this problem is Boerhaave who, in experiments with a mass of iron weighing 5 lb. 8 oz. when it was cold, found that its weight both when it was red-hot and after it had cooled was the same as when it was cold (*Elementa Chemiæ*, Leyden, 1732, I, pp. 259–60: English trans., *A New Method of Chemistry*, by P. Shaw, London, 1741, I, pp. 285–6). Buffon,[2] on the other hand, found that a mass of iron weighing 49 lb. 9 oz. when "white-hot" weighed 2 oz. less when cooled to atmospheric temperature, which at the time of his experiments was at freezing-point; and that other masses of iron gave similar results (*Histoire Naturelle, Supplément*, Paris, 1775, II, pp. 11–13).

Roebuck,[3] using accurate balances[4] to check Buffon's results,

[1] See McKie, *Science Progress*, 1934, **29**, 253.

[2] Jean Louis Le Clerc, Comte de Buffon (1707–88), an extensive writer on scientific subjects and author of the famous *Histoire Naturelle* (Paris, 44 vols., 1749–1804).

[3] John Roebuck, M.D. (1718–94), chemist and physician, pupil of Cullen and Black, friend and sometime partner of James Watt. Founder of the Carron ironworks. Introduced lead-chambers in place of glass vessels for the manufacture of sulphuric acid and thus reduced the price to one-quarter of what it was before. As one of the earliest consulting chemists, Roebuck played a great part in the establishment of many industries at Birmingham.

[4] One of these balances would weigh masses of 1 lb. weight and turn with $\frac{1}{10}$ of a grain, and the other would weigh masses of half-ounce weight and turn with $\frac{1}{100}$ of a grain.

which he thought doubtful, found in repeated experiments that a mass of iron weighing nearly 1 lb. and heated to a white heat "or what the smiths call a welding heat" weighed "nearly one grain less when cold than when hot," but a smaller mass, about five pennyweights, weighed "somewhat more when cold than hot," while a piece of copper, weighing nearly 1 lb. when hot, was 4 grains lighter when cool (*Phil. Trans.*, 1776, **66**, 509). In the case of copper Roebuck further showed that the decrease in its weight on cooling was almost entirely accounted for by the weight of the scales that fell from it.   He then heated a cylinder of wrought iron weighing 55 lb. to white-heat and counterpoised it on a balance: after cooling for 2, 5, 6, and 22 hours, it gained in weight 3 pennyweights and a few grains, 3 pennyweights and 17 grains, 4 pennyweights, and 6 pennyweights and 17 grains respectively.   To find the cause of this increase, Roebuck then heated 2 oz. 8 pennyweights of the scales of wrought iron and found that they increased in weight by 5 grains when cold: and similarly two pieces of pure silver, weighing 2 lb. 10 oz. 5 pennyweights when hot, showed when cold an increase of weight amounting to 5 grains, although no calx was formed.   Roebuck's experiments thus proved indecisive.

Whitehurst [1] (*Phil. Trans.*, 1776, **66**, 575) obtained results disagreeing with those of Buffon.   A pennyweight of gold made red-hot became lighter, but when cooled to its original temperature (atmospheric temperature) "its former weight was perfectly restored."   An equal mass of iron treated in the same way was also apparently reduced in weight, but when cooled "its weight was visibly augmented."   Whitehurst's experiments had been made several years before publication and he had repeated them many times.   The beam of his balance was sensibly affected by $\frac{1}{2000}$th of a grain.   The metals were heated in charcoal by means of a candle and blow-pipe, and both were

---

[1] John Whitehurst (1713–88), scientific instrument-maker at Derby. Appointed stamper of the money-weights at London in 1775 on the passage of the Act for the better regulation of the gold coinage. Author of a number of geological works.

brought nearly to the point of fusion. Whitehurst ascribed the "apparent levity" of the gold and iron, when hot, to the ascent of rarefied air above the scales "and to the tendency of that underneath to restore the equilibrium of its pressure." The iron, he noted, might have increased in weight from partial conversion to steel through the action of the flame and the charcoal. He was at a loss to account for the fallacy of Buffon's results, but suggested "that the heat of the mass of iron employed by him had a greater effect on that arm of the beam from which it hung than on the other, which being less heated, would consequently be less expanded; and this difference of expansion might produce the error in M. Buffon's account of the weight of heated iron."

The advances made by Black now began to influence the treatment of this problem. Bergman [1] (*De Attractionibus Electivis* in *Opuscula*, Upsala, 1783, III, p. 427; English trans., *Dissertation on Elective Attractions*, London, 1785, pp. 244–5) argued that, since the "subtile matter" of heat doubtless gravitated, it would be more satisfactory to attempt to weigh it in fluids where it was more abundant, than in solids where it appeared to elude the accuracy of the instruments employed. Later he added:

Water heated to 130 degrees melts an equal weight of snow; but the water thus brought to a liquid state is at the point of congelation. It would certainly be worth while to weigh, with the utmost exactness, a piece of ice in a perfectly close vessel, and to repeat the operation after it was melted. The stopple must fit, in the most accurate manner, lest any thing should be lost in consequence of evaporation. This experiment has not yet, as far as I know, been performed with proper care and accuracy [2]; it may, however, ascertain, in some

---

[1] Tobern Olaf Bergman (1735–84), the famous chemist. Professor of Chemistry in the University of Upsala, 1767–84.

[2] More than a hundred years before this, Boyle, in seeking to detect the alleged "Corpuscles of Cold," had weighed water, frozen and unfrozen, enclosed either in a phial "nimbly clos'd with *Hermes*'s seal" or in a "Pewter Box" and found "not one grain difference" in its weight (*New Experiments and Observations Touching Cold*, London, 1665, pp. 556–9 and 572–3).

measure, the absolute weight of a quantity of the principle of heat corresponding to 130 degrees (*ibid.*, English trans., p. 257).[1]

Accordingly we find that Fordyce [2] (*Phil. Trans.*, 1785, **75**, 361), when announcing that he had found in numerous experiments that bodies lost in weight on being melted or heated, reported only one, namely, the loss of weight when ice melted into water. He took a glass globe, weighing about 451 grains and containing about 1700 grains of New River water. This was hermetically sealed and the whole cleaned and weighed on a balance in which $\frac{1}{1600}$th part of a grain caused a movement of one division of the index when there was a load of 4 to 5 oz. in each scale-pan. It weighed $2150\frac{31}{32}$ grains at $32°$ F., Fordyce having placed it in a freezing mixture of ice and salt and then removed it just as it began to freeze. It was then put back in the freezing mixture for 20 minutes. Partial freezing occurred and, when taken out, dried and reweighed, its weight was found to have increased by about $\frac{1}{60}$th of a grain. This was repeated five times; each time more water had frozen and more weight was gained. When it was completely frozen, it was found to have gained "$\frac{3}{16}$ of a grain and four divisions of the index." Left to stand in the scale-pan for about a minute, it began to lose weight. Fordyce removed it to a distance and, inserting a thermometer in the freezing mixture, found that the

---

[1] This passage, except for the first sentence, occurs only in the English version. The *Elective Attractions* was first published in 1775 (*Nova Acta Reg. Soc. Sci. Upsal.*, 1775, **2**, 159), an abbreviated French translation appearing in Rozier's Journal in 1778 (*Observations etc.*, 1778, **13**, Supplément, 298). Bergman extended the work later (*Opuscula Physica et Chemica*, Upsala, 1783, III, p. 291), including among other additions a section on the "matter of heat," and it was apparently from this text that the English translation was made, although the passage we have quoted above does not occur in the original and is evidently an interpolation. The "130 degrees" in the first sentence corresponds to Wilcke's "72 degrees" (in the original), owing to the change from the Centigrade to the Fahrenheit scale.

[2] George Fordyce (1736–1802), pupil of Cullen. Physician at St. Thomas's Hospital, London, 1770–82. Wrote a number of medical works. Earned a great reputation in London from 1759 onwards for the private lectures he gave to medical students before the organization of regular medical schools.

temperature had fallen to 10° F. while the temperature of the glass globe containing the ice was 12° F.   It was then left for half an hour, at the end of which time the temperature of the globe was 32° F.   It was then reweighed: and "it had lost $\frac{1}{8}$ and five divisions; so that it weighed $\frac{1}{16}$, all but one division, more than when the water was fluid."

The ice was then almost completely melted, except for a small piece, and exposed to the air at 32° F. for a quarter of an hour. Reweighed, it proved "heavier than the water was at first one division of the beam."   The weights being removed, the beam was exactly balanced as before the experiment.

This increase of weight, occurring when water was converted into ice, might, said Fordyce, be due either to increased gravitation or to the entry of extraneous matter through the glass.   To decide which of these was the case Fordyce suggested "forming a pendulum of water, and another of ice, of the same length, and in every other respect similar, and making them swing equal arcs.   If they mark equal times, then certainly there is some matter added to the water.   If the pendulum of ice is quicker in its vibrations, then the attraction of gravitation is increased."   Until this or some similar experiment was done, the cause of the increase of weight observed could merely be conjectured.

Fordyce argued too that the increase in weight could not be explained by the contraction of the glass vessel at the low temperatures and the cooling of the air above the scale-pan containing it, but added that he was constructing an apparatus "in which this cause of fallacy will be totally removed." [1]   At the moment, all that could be said was "that water gains weight on being frozen."

This problem now attracted one of the leading scientists of the time, Benjamin Thompson, Count Rumford,[2] who is perhaps better known for his classic work on the heat developed in the boring of guns from which he argued that heat was a form

[1] We cannot find any record of this apparatus.

[2] Sir Benjamin Thompson, Count Rumford (1753–1814).   F.R.S.ʼ and one of the founders of the Royal Institution.

of motion (*Phil. Trans.*, 1798, **88**, 80). Rumford, much interested in Fordyce's results and possessed of an excellent balance, set out to repeat these experiments in 1787, but did not publish his results until twelve years afterwards (*Phil. Trans.*, 1799, **89**, 179). This work and Rumford's deductions constitute the most serious attack made on the material theory of heat before the middle of the nineteenth century.

Rumford selected two thin-walled Florence flasks of the same shape and size and which, by every possible method of comparison, appeared indistinguishable from one another. One of these, A, was half-filled with 4107·86 grains of pure distilled water; the other, B, contained an equal mass of weak spirits of wine. Both were hermetically sealed, washed, and wiped "perfectly clean and dry on the outside." They were then suspended one from each arm of a balance and left in that position to acquire the temperature of the room, 61° F., maintained constant except for "very little variation." When Rumford thought that they had acquired this temperature, he wiped them again "with a very clean dry cambric handkerchief" and balanced them exactly by affixing a small piece of very fine silver wire to the arm of the balance from which the lighter bottle was suspended. Left in this state for twelve hours, they showed no alteration in their relative weight. Rumford now moved the whole apparatus into a room, facing north and at a temperature of 29° F., the atmospheric temperature at this time being 27° F. After 48 hours he "found that the bottle A very sensibly preponderated." The water in it was completely frozen, while the spirits of wine in the other bottle, B, were quite fluid. He re-adjusted the equilibrium of the balance by adding small pieces of gold-lace wire, and thus found that the bottle A had increased in weight by the $\frac{1}{35904}$th part of its original weight,[1] i.e. $\frac{134}{1000}$ parts of a grain had to be added to the opposite arm of the balance to re-counterbalance. This was "very contrary" to Rumford's expectation. He returned the balance and the bottles to the warmer room (at 61° F.); the ice

---

[1] 4811·23 grains, the bottle weighing 703·37 and the water 4107·86 grains.

thawed gradually, and he presently found, when the ice was all thawed and when the water thus produced had acquired the temperature of the room, that both bottles, wiped perfectly clean and dry, weighed the same as at the beginning of the experiment.

Rumford repeated the experiment; and the water, as before, appeared to be heavier when frozen, "but some irregularity," he said, "in the manner in which the water lost the additional weight which it had appeared to acquire upon being frozen, when it was afterwards thawed, as also a sensible difference in the quantities of weight apparently acquired in the different experiments, led me to suspect, that the experiment could not be depended on for deciding the fact in question."   He suspected the accuracy of the balance.   He therefore took two equal solid brass globes, "well gilt and burnished," weighing 4975 grains each, and suspended one from each arm of the balance by fine gold wires.   He had previously found that "the gilt surfaces of metals" did not attract moisture: and he was afraid that despite all his precautions moisture had been deposited on the glass globes in the previous experiments.   The brass globes and the balance after standing for 24 hours in the room at 61° F. were carefully balanced and then removed to a room at 26° F., where they were left all night.   "The result of this trial," said Rumford, "furnished the most satisfactory proof of the accuracy of the balance; for, upon entering the room, I found the equilibrium as perfect as at the beginning of the experiment."   The balance was, therefore, accurate.   Rumford, moreover, thought that he had carefully guarded himself against all possible errors in the freezing experiments, especially against a possible difference in the amounts of moisture condensed on the glass flasks. He now proceeded by admitting Fordyce's result, which his own experiment tended to confirm, and trying to find an explanation of it.   And the only explanation that he could hit upon was that the water in freezing had lost a great proportion of its latent heat, and "if the loss of latent heat added to the weight of one body, it must produce the same effect on another, and consequently, that the augmentation of the quantity of latent heat must—in all bodies—and in all cases—diminish their apparent weights."

Rumford proceeded to test this, taking two bottles "as nearly alike as possible" and similar to those previously used, placing 4012·46 grains of water in one of them and an equal mass of mercury in the other, sealing them hermetically, suspending one from each arm of the balance and leaving them to acquire the temperature of the room, 61° F. This done, he balanced them exactly and then removed them with the balance to another room at 34° F. where they stood 24 hours. "But," he reported, "there was not the least appearance of either of them acquiring, or losing, any weight." Now, argued Rumford, the heat lost by the water must have exceeded greatly that lost by the mercury, since the specific heats of water and mercury were known to be as 1000 to 33: yet this difference in the amounts of heat lost "produced no sensible difference on the weights of the fluids in question." Moreover, his balance would have detected a change in weight of one-millionth part of the weight of either fluid, and he could measure a change of $\frac{1}{700000}$ part of that weight.

Rumford therefore concluded that the apparent increase of weight of water on freezing that occurred in his first experiments was due to some "accidental cause," which he could not identify, but which was, he suspected, either unequal deposition of atmospheric moisture on the surfaces of the bottles or vertical currents of air produced by one or both of the bottles not being exactly at the same temperature as the surrounding atmosphere.

He therefore tried to set the matter beyond all doubt by means of a further experiment. He took three bottles, A, B and C, as before. Into A he put 4214·28 grains of water and a small thermometer with its bulb arranged centrally in the water; into B he put an equal mass of spirits of wine and a similar thermometer; and into C he put an equal mass of mercury. These bottles, all hermetically sealed, were left in the room at 61° F. for 24 hours. The thermometers then indicated that the contents of A and B were exactly at the same temperature, and all the bottles were then carefully wiped free from all moisture on

K

the outside and left for 2 hours longer in case the thermal equilibrium had been thereby disturbed. They were then all weighed and exactly balanced with pieces of very fine silver wire attached to the necks of the lighter bottles. This done, they were all removed to a room at 30° F. and left there undisturbed for 48 hours, A and B attached to the arms of the balance and C suspended at an equal height from a suitable stand placed as near the balance as possible and carrying a sensitive thermometer. At the end of this time, the three thermometers all read alike, 29° F., that in A being now enclosed in the ice produced by the freezing of the water; and A and B were found "remaining in the most perfect equilibrium." The beam of the balance being slightly and gently displaced moved quite freely and then, when vibration ceased, came to rest precisely at its former position. Rumford then removed bottle B and replaced it with C, and this also showed that its weight was unchanged, "being in the same perfect equilibrium with the bottle A as at first."

The whole apparatus was now moved to a warm room, the ice in A melting and the bottles being allowed to attain the temperature of the room; and their weights, when they were wiped clean and dry, were found to be unchanged. The experiment was repeated several times, always with the same result, "the water, *in no instance*, appearing to gain, or to lose, the least weight, upon being frozen, or upon being thawed," and the relative weights of the spirits of wine and the mercury remaining constant at the various temperatures to which they were exposed.

The differences observed in the first experiments Rumford now ascribed to the possibility that the contents of the bottles might not have been precisely at the same temperature, which inequality might have produced convection currents in the air immediately surrounding them or might have led to unequal quantities of moisture being deposited on the surfaces of the bottles "or to both these causes operating together."

Having determined [concluded Rumford] that water does not acquire or lose any weight, upon being changed from a state of *fluidity* to that of *ice*, and *vice versa*, I shall now take my final leave of a subject which has long occupied me, and which has cost me much pains and trouble; being fully convinced (from the results of the above-mentioned experiments) that if heat be in fact a *substance*, or matter —a fluid *sui generis*, as has been supposed—which, passing from one body to another, and being accumulated, is the immediate cause of the phenomena we observe in heated bodies, (of which, however, I cannot help entertaining doubts,) it must be something so infinitely rare, even in its most condensed state, as to baffle all our attempts to discover its gravity.   And, if the opinion, which has been adopted by many of our ablest philosophers, that heat is nothing more than an intestine vibratory motion of the constituent parts of heated bodies, should be well founded, it is clear that the weights of bodies can in no wise be affected by such motion.

It is, no doubt, upon the supposition that heat is a substance distinct from the heated body, and which is accumulated in it, that all the experiments which have been undertaken, with a view to determine the weight which bodies have been supposed to gain, or to lose, upon being heated or cooled, have been made; and, upon this supposition (but without, however, adopting it entirely, as I do not conceive it to be sufficiently proved,) all my researches have been directed.

Further, added Rumford, the experiments with water and ice were the most favourable that could be contrived for detecting the weight of heat, since the quantity of heat lost by water in freezing was so considerable, namely, 140 degrees of Fahrenheit's scale, which amount if added to water at 32° F. would bring it within 60 [1] degrees of its boiling-point, and, since his own experiments employed water at 61° F., the difference in the quantities of heat contained by the water at this temperature and by the ice at 32° was nearly equal to that between water at its B.P. and water at its F.P.   The amount of heat lost by water in freezing appeared, he said, even more considerable if it was realized that, since its specific heat was to that of gold as 1000 to 50, or as 20 to 1, it would, if communicated to an equal mass of gold at 32° F. raise the temperature of the gold to 2800° (i.e. 140 × 20), that is, to a bright red heat.

[1] As in original, but it should read 50, since 32 + 140 = 172, this being given as 162 in original.   The correct figure is 40.

It appears therefore [ended Rumford] to be clearly proved, by my experiments, that a quantity of heat equal to that which 4214 grains (or about 9¾ oz.) of gold would require to heat it from the temperature of freezing water to be *red hot*, has no sensible effect upon a balance capable of indicating so small a variation of weight as that of $\frac{1}{1000000}$ part of the body in question; and, if the weight of gold is neither augmented nor lessened by *one millionth part*, upon being heated from the point of *freezing water* to that of *a bright red heat*, I think we may safely conclude, that ALL ATTEMPTS TO DISCOVER ANY EFFECT OF HEAT UPON THE APPARENT WEIGHTS OF BODIES WILL BE FRUITLESS.

Rumford's concluding sentence was set up in this large type: and so began, and so for the moment ended, the attack on the material theory of heat in a memoir that describes a magnificent experimental technique and stands out as a classic example of what scientific investigation is at its best.   Yet, as is well known, the attack failed: the rejection of the material theory was still more than half a century away.

APPENDIX

Johannis Baptistae Morini, *Astrologia Gallica*,
Lib. VIII, Cap. XV, pp. 158–9

Quommodo invenienda sit temperies duorum miscibilium ejusdem consistentiae, sed diversae temperiei, si invicem misceantur et alterentur. Hac de re (quod sciam) nemo hactenus quidquam determinavit: Hoc autem scire nec injucundum, nec forsan inutile fuerit, praesertim Medicis: itaque quandoquidem Philosophorum communis est sententia mixtum admittere duntaxat 8 gradus contrarium qualitatum, hoc supposito. Si aqua calida ut duo, frigida autem ut sex, misceatur cum calida ut quatuor, frigida autem ut quatuor; neque fiet mixtum calidum ut duo, neque mixtum calidum ut quatuor; non primum, quia sequeretur a frigido, ut sex quod dicitur calidum ut duo, corrumpi duos gradus caloris in calido ut quatuor, non reagente calido ut quatuor in frigidum ut sex, quod est absurdum; cum in frigido ut sex, sit potentia recipiendi caloris, et in calido ut quatuor potentia producendi caloris: nec non in mixtione debite applicentur agens et patiens in qualitatibus dissimiles. Non secundum, alias calidum ut quatuor producet in frigido ut sex duos gradus caloris non reagente frigido, quod est idem absurdum. Ergo neque fiet mixtum calidum ut duo, neque calidum ut quatuor; fiet tamen mixtum plures habens quam duos gradus caloris, quia calidum ut quatuor reaget in frigidum ut sex; attamen pauciores quam quatuor gradus caloris, quia frigidum sex reaget in calidum ut quatuor: Neque fiet mixtum calidum ut tria, alioquin sequeretur frigidum ut sex et calidum ut quatuor, in eodem temporis spatio producere effectus aequales seu ejusdem intensionis; quod ideo absurdum est, quia sex gradus frigoris fortiores sunt quatuor gradibus caloris, et frigidum ut sex est similius frigido ut quatuor, quam calidum ut quatuor calido ut duo, et illud majorem habet resistentiam quam hoc, ergo eodem tempore non producent effectum aequalis intensionis. Ut vero gradus ex mixtione resultans determinetur.

Nota primo, quod in mixtione nulla fit qualitatum miscibilium destructio aut corruptio; sed tantum miscella; calido et frigido suas qualitates in se mutuo effundentibus. Si autem unum sic intendatur, ut cum altero sit incompatibile, tunc expellitur ipsum alterum.

Nota secundo, quod duo gradus caloris existentes in frigido ut sex; et quatuor gradus frigoris existentes in calido ut quatuor, nihil agunt vel reagunt in hac mixtione. Non enim calidum ut duo, agit in frigidum ut sex, cum quo est; alioquin se etiam intenderet, sex gradus frigoris remittendo: sed neque etiam dum fit mixtio potest agere in frigidum, ut quatuor sociatum calido ut quatuor; alioquin frigidum ut quatuor remittendo, intenderet calidum ut quatuor, quod est supra vires ejus, ideoque absurdum. Idem dicito de frigido ut quatuor. Ergo in mixtione nihil agunt, sed tantum a sibi similibus intenduntur. Nec dicas si intenduntur pati, ideoque re agere: quia cum intendantur a suis similibus, sed intensioribus, neque possunt, neque debent in ea re agere. Hincque liquido constat, actionem et reactionem in mixtione reperiri tantum inter intensiora contraria, intensionem vero inter remissiora, quod est bene notandum.

Nota tertio. Totalem virtutem sex graduum frigoris, scilicet tam agendi quam resistendi, aequalem esse totali virtuti sex graduum caloris; licet enim activitas sex graduum caloris sit major quam activitas sex graduum frigoris, ex eo tamen consurgit semper aequaliter inter totales ipsas virtutes, quod tantum resistentia caloris superatur a resistentia frigoris, quantum activitas frigoris ab activitate caloris: quae enim qualitates sunt magis activae, eae sunt minus resistivae: Totalis ergo virtus sex graduum frigoris dupla erit totalis virtutis trium graduum caloris: et totalis virtus sex graduum frigoris sesqui altera totalis virtutis quatuor graduum caloris.

His positis ad propositum dico. Quod si frigidum ut sex misceatur cum calido ut quatuor, frigus productum in calido, eam habebit rationem ad calorem productum in frigido, quam frigidum habet ad calidum: ac proinde cum eorum virtutes totales sint sesqui alterae, effectus inde nascentes erunt sesqui alteri.

Proinde ut sciatur quantum frigoris producat in calido frigidum, et quantum caloris in frigido calidum, nec non gradus temperiei mixtionis, sic per Algebram operare.

Quaeruntur duo numeri in proportione sesqui altera, quorum major representans frigus productum, additus quatuor gradibus frigoris qui cum calido ut quatuor reperiuntur; minor autem exprimens gradus caloris producti, sublatus a sex gradibus frigoris: summa et residuum sint aequalia.

Sit major 3A, minor 2A, majori adde 4 et minorem aufer a 6, habebit 3A + 4 aequales 6 — 2A: Da utrique 2A: habebis 5A + 4 aequales 6. aufer ab utroque 4, habebis 5A aequales 2. Deinde 2 per 5, habebis $\frac{2}{5}$ pro valore 1A, quare 3A valebunt $\frac{6}{5}$, et 2A valebunt $\frac{4}{5}$, qui sunt numeri quaesiti. Ergo frigidum in calido producet $\frac{6}{5}$ frigoris: Et calidum in frigido $\frac{4}{5}$ caloris. Ut vero gradus temperiei mixtionis sciatur, adde $\frac{6}{5}$ frigoris cum quatuor gradibus frigoris, habebit in mixtione $5\frac{1}{5}$ gradus frigoris: quibus sublatis ab 8 gradibus contrarietatis, quanta est materiae latitudo, seu capacitas; supersunt gradus $2\frac{4}{5}$ caloris: Mixtum ergo habebit $5\frac{1}{5}$ gradus frigoris, et $2\frac{4}{5}$ gradus caloris, sic de caeteris. Porro superiora sunt intelligenda cum miscibilia sunt aequalia in materiae quantitate. Nam si V.C. materia frigidi ut sex dupla sit calidi ut quatuor, duplanda erit frigidi virtus totalis: quia in duplo materiae, duplum est virtutis: si tripla triplanda: Tum cum proportione virtutum totalium calidi et frigidi reliqua ut supra peragenda: sic de reliquis materiae proportionibus. Atq; ideo si calidum ut sex miscendum sit cum frigido ut septem. Si autem calidi materia sesqui altera ad frigidi materiam, nascitur mixtum calidum in $3\frac{13}{16}$ grad. et frigid. in $4\frac{3}{16}$ grad.

INDEX

HISTORY, PHILOSOPHY AND SOCIOLOGY OF SCIENCE

Classics, Staples and Precursors

An Arno Press Collection

Aliotta, [Antonio]. **The Idealistic Reaction Against Science.** 1914

Arago, [Dominique François Jean]. **Historical Eloge of James Watt.** 1839

Bavink, Bernhard. **The Natural Sciences.** 1932

Benjamin, Park. **A History of Electricity.** 1898

Bennett, Jesse Lee. **The Diffusion of Science.** 1942

[Bronfenbrenner], Ornstein, Martha. **The Role of Scientific Societies in the Seventeenth Century.** 1928

Bush, Vannevar. **Endless Horizons.** 1946

Campanella, Thomas. **The Defense of Galileo.** 1937

Carmichael, R. D. **The Logic of Discovery.** 1930

Caullery, Maurice. **French Science and its Principal Discoveries Since the Seventeenth Century.** [1934]

Caullery, Maurice. **Universities and Scientific Life in the United States.** 1922

Debates on the Decline of Science. 1975

de Beer, G. R. **Sir Hans Sloane and the British Museum.** 1953

Dissertations on the Progress of Knowledge. [1824]. 2 vols. in one

Euler, [Leonard]. **Letters of Euler.** 1833. 2 vols. in one

Flint, Robert. **Philosophy as Scientia Scientiarum and a History of Classifications of the Sciences.** 1904

Forke, Alfred. **The World-Conception of the Chinese.** 1925

Frank, Philipp. **Modern Science and its Philosophy.** 1949

The Freedom of Science. 1975

George, William H. **The Scientist in Action.** 1936

Goodfield, G. J. **The Growth of Scientific Physiology.** 1960

Graves, Robert Perceval. **Life of Sir William Rowan Hamilton.** 3 vols. 1882

Haldane, J. B. S. **Science and Everyday Life.** 1940

Hall, Daniel, et al. **The Frustration of Science.** 1935

Halley, Edmond. **Correspondence and Papers of Edmond Halley.** 1932

Jones, Bence. **The Royal Institution.** 1871

Kaplan, Norman. **Science and Society.** 1965

Levy, H. **The Universe of Science.** 1933

Marchant, James. **Alfred Russel Wallace.** 1916

McKie, Douglas and Niels H. de V. Heathcote. **The Discovery of Specific and Latent Heats.** 1935

Montagu, M. F. Ashley. **Studies and Essays in the History of Science and Learning.** [1944]

Morgan, John. **A Discourse Upon the Institution of Medical Schools in America.** 1765

Mottelay, Paul Fleury. **Bibliographical History of Electricity and Magnetism Chronologically Arranged.** 1922

Muir, M. M. Pattison. **A History of Chemical Theories and Laws.** 1907

National Council of American-Soviet Friendship. **Science in Soviet Russia: Papers Presented at Congress of American-Soviet Friendship.** 1944

Needham, Joseph. **A History of Embryology.** 1959

Needham, Joseph and Walter Pagel. **Background to Modern Science.** 1940

Osborn, Henry Fairfield. **From the Greeks to Darwin.** 1929

Partington, J[ames] R[iddick]. **Origins and Development of Applied Chemistry.** 1935

Polanyi, M[ichael]. **The Contempt of Freedom.** 1940

Priestley, Joseph. **Disquisitions Relating to Matter and Spirit.** 1777

Ray, John. **The Correspondence of John Ray.** 1848

Richet, Charles. **The Natural History of a Savant.** 1927

Schuster, Arthur. **The Progress of Physics During 33 Years (1875-1908).** 1911

Science, Internationalism and War. 1975

Selye, Hans. **From Dream to Discovery: On Being a Scientist.** 1964

Singer, Charles. **Studies in the History and Method of Science.** 1917/1921. 2 vols. in one

Smith, Edward. **The Life of Sir Joseph Banks.** 1911

Snow, A. J. **Matter and Gravity in Newton's Physical Philosophy.** 1926

Somerville, Mary. **On the Connexion of the Physical Sciences.** 1846

Thomson, J. J. **Recollections and Reflections.** 1936

Thomson, Thomas. **The History of Chemistry.** 1830/31

Underwood, E. Ashworth. **Science, Medicine and History.** 2 vols. 1953

Visher, Stephen Sargent. **Scientists Starred 1903-1943 in American Men of Science.** 1947

Von Humboldt, Alexander. **Views of Nature: Or Contemplations on the Sublime Phenomena of Creation.** 1850

Von Meyer, Ernst. **A History of Chemistry from Earliest Times to the Present Day.** 1891

Walker, Helen M. **Studies in the History of Statistical Method.** 1929

Watson, David Lindsay. **Scientists Are Human.** 1938

Weld, Charles Richard. **A History of the Royal Society.** 1848. 2 vols. in one

Wilson, George. **The Life of the Honorable Henry Cavendish.** 1851